AF614737

OZONE IN NATURE AND PRACTICE

Edited by **Ján Derco** and **Marian Koman**

Ozone in Nature and Practice
http://dx.doi.org/10.5772/intechopen.68925
Edited by Ján Derco and Marian Koman

Contributors

Aysan Lektemur Alpan, Olcay Bakar, Tatyana Poznyak, Arizbeth Pérez, Clara-Leticia Santos Cuevas, Pamela Guerra, Isaac Chairez, Julia Liliana Rodríguez, Iliana Fuentes, C. Marissa Aguilar, Miguel A. Valenzuela, Jan Derco

First published in London, United Kingdom, 2018 by IntechOpen
IntechOpen is the global imprint of INTECHOPEN LIMITED, registered in England and Wales, registration number: 11086078, The Shard, 25th floor, 32 London Bridge Street
London, SE19SG – United Kingdom
Printed in Croatia

British Library Cataloguing-in-Publication Data
A catalogue record for this book is available from the British Library

Additional hard copies can be obtained from orders@intechopen.com

Ozone in Nature and Practice, Edited by Ján Derco and Marian Koman
p. cm.
Print ISBN 978-1-78923-382-7
Online ISBN 978-1-78923-383-4

For EU product safety concerns:
IN TECH d.o.o., Prolaz Marije Krucifikse Kozulić 3, 51000 Rijeka, Croatia,
info@intechopen.com or visit our website at intechopen.com.

Meet the editors

Prof. Ján Derco, DSc, graduated from the Faculty of Chemical and Food Technology, Slovak University of Technology (SUT) with an MSc degree in Chemical Engineering. He then started working at the Department of Environmental Engineering, Faculty of Chemical and Food Technology, where he has continued his research. Later he obtained his PhD and DSc (doctor of sciences) graduations from the same university. The main fields of his scientific interest are environmental engineering, biological wastewater treatment, modeling, design and optimization, ozone-based oxidation processes, priority substances, micropollutant degradation and transformation. He is presently working as a Professor at the Institute of Chemical and Environmental Engineering, SUT.

Prof. Marian Koman, DSc, graduated from the Faculty of Chemical and Food Technology, Slovak University of Technology (SUT) with an MSc degree in Chemical Engineering. He then started working at the Department of Inorganic Chemistry, Faculty of Chemical and Food Technology, where he has continued his research. Later he obtained his PhD and DSc (Doctor of Sciences) graduations from the same university. The main fields of his scientific interest are inorganic chemistry, coordination chemistry (including the preparation of complexes of Fe and Mn for catalytic ozonation processes) and X-ray structure analysis. He is presently working as a Professor at the Institute of Inorganic Chemistry, Technology and Materials, SUT.

Contents

Chapter 1

Introductory Chapter: Ozone in the Nature and Practice

Ján Derco, Barbora Urminská and Martin Vrabeľ

Additional information is available at the end of the chapter

http://dx.doi.org/10.5772/intechopen.78263

1. Introduction

Ozone (O_3) is a tri-atomic form of oxygen, that is, three atoms of oxygen bonded together. It is an unstable gas with distinctive sharp odor. Under normal conditions, ozone is unstable and rapidly decomposes to a more stable gaseous form of molecular oxygen, O_2. Behavior of ozone shows more "faces" according to the circumstances. The ozone layer protects the Earth from damaging ultraviolet (UV) light; therefore, it is an inevitable compound to save life on Earth. Its applications also present benefits for human welfare. Thanks to its strong oxidizing properties as well as bactericidal, virucidal, and fungicidal effects, it is widely used for water disinfection, wastewater treatment, in various industrial fields, for example, in the pulp and paper industry and the textile industry and progressively for medical purposes as well. Unfortunately, there are also circumstances, mainly due to the activities of humans, when ozone behaves contrary to the mentioned benefits, with negative and harmful influence on human health, and life in nature, as well as to material environment [1]. While stratospheric ozone protects biological life on Earth, ozone in the troposphere is toxic to human health, plants, and trees and it also damages various materials [2].

One may have noticed a smell in the air after thunders and lightning bolts. The pure odor was most likely to the actual ozone formed by flashing atmospheres. Because ozone is unstable and cannot be stored successfully, it must be generated at the application site. To produce ozone sufficiently in the water, wastewater, and sewage treatment plant, ozone generators are used. The simplest method can be ozone generated by the passage of oxygen or oxygen-containing air through the area of electric discharge or spark.

Ozone is a subject of interest and research for a number of reasons. This chapter will deal briefly with related aspects of stratospheric and tropospheric ozone as well as with utilization of ozone in practice.

2. Ozone in the environment

In the case of atmospheric ozone, the main focus is on the decrease in the concentration of stratospheric ozone and the related biological effect, the occurrence, and the detrimental effects of tropospheric ozone.

Ozone is highly toxic, blue to colorless gas with a pungent odor. It is located mainly in the stratosphere (15–50 km above the Earth's surface) and the troposphere (0–11 km above Earth's surface).

2.1. Tropospheric ozone

The most important chemical transformation processes in the atmosphere are photo-dissociation and oxidation. Major atmospheric oxidants include O_3, hydroxyl radical ·OH, and hydrogen peroxide H_2O_2 (together they form the oxidative capacity of the atmosphere). Among possible consequences of transformation, there is a production of substances that are more toxic than their precursors (i.e. the substances leading to their formation). An example is the formation of tropospheric ozone, so-called ground-level ozone, which is harmful and is among the toxic pollutants. Ground-based ozone is one of the most variable atmospheric components that depend on temperature and air pressure [2].

Ground-level ozone is released when various hydrocarbons react with nitrogen dioxide (emissions from public transportation and industry) in the presence of sunlight. The largest source of nitrogen oxides (NO_x) is the combustion of fossil fuels (energy, transport) and therefore the largest share of automotive transport. Exhaust gases from engines include NO_x (approximately 95% NO, 5% NO_2), hydrocarbons (saturated and unsaturated), partially oxidized hydrocarbons (mainly aldehydes, most of them formaldehyde), and CO. Levels of ground-level ozone in the exterior are higher during the summer [3, 4].

High concentrations of ozone adversely affect human health. Ozone as an effective oxidant causes transient irritation of the respiratory system, resulting in coughing, irritation of the nose, throat, eyes, breathing, and chest pain in deep breathing. Chronic exposure to high concentrations of ozone leads to premature aging of the lung tissue, thereby reducing resistance to infections. Up to 95% of ozone inhaled into the lungs remains in the body.

Tropospheric ozone also has an unfavorable effect on vegetation, especially on agricultural crops and forest stands. Smog-induced foliar injury on plants was first identified in the 1950s [2]. Short episodes of high concentrations of ground-level ozone cause acute damage to assimilation organs (reduction of resistance or decay of the plant).

Ground-level ozone effectively decomposes chlorophyll in plants. It also damages various materials. It causes rubber hardening, decoloration of dyes and dyed textiles, and corrosion of metals. Increases in corrosion of construction materials (steel, zinc, copper, aluminum, and bronze) are attributed to the synergistic action of ozone and other substances such as SO_2 and NO_2 [2].

The World Health Organization (WHO) recommends an 8-h limit of 100–120 μg m^{-3} concentration of ozone-exposed to human body [1].

2.2. Stratospheric ozone

Ozone is formed in the stratosphere by the reaction of an oxygen radical with an oxygen molecule. The ozone molecule can absorb radiation in the range of 310–200 nm, after which it is divided into an oxygen molecule and an oxygen radical, followed by a repeat cycle [3].

$$O_2 + UV_{240\ nm} \rightarrow \cdot O + \cdot O \cdot O + O_2 \rightarrow O_3 \tag{1}$$

$$O_3 + UV_{310\text{–}200\ nm} \rightarrow \cdot O + O_2 \tag{2}$$

It can be destroyed by a number of free radicals, the largest of which are: $\cdot OH \cdot NO$, $\cdot Cl$ and $\cdot Br$. While ·OH and ·NO radicals are mostly of natural origin, the halogen radicals are increased due to the human activity by discharging chlorine and fluorine derivatives of hydrocarbons.

$$\text{(CFC, HCFC)}\ CFCl_3 + UV \rightarrow \cdot C + CFCl_2 \tag{3}$$

Subsequently, the radicals act as a catalyst.

$$\cdot C + O_3 \rightarrow ClO + O_2\ ClO + O_3 \rightarrow \cdot Cl + 2\ O_2 \tag{4}$$

2.3. Ozone hole

Ozone hole is a term describing more than 50% temporary loss of ozone in the stratosphere. At poles, the ozone layer is already very thin. The most significant decrease in ozone was recorded in the lower stratospheric levels a decrease up to 70% was recorded over the Antarctic and has continued since 1985. The restoration was recorded in 2016 similarly to the Arctic. The reason why ozone holes are formed more easily over the poles is due to the phenomenon known as polar vortex that allows chemical reactions in the enclosed air mass to be enhanced due to the lack of mixing with other, lower latitude, air masses. The effects of the pollutants in the atmosphere are thus enhanced in these isolated regions of the atmosphere [5]. Another reason is the formation of polar stratospheric clouds that are composed of ice crystals, sometimes greatly enriched in nitrogen oxides enhancing the ozone degradation. During winter, Antarctic is extremely cold due to the absence of sunlight. The temperature here drops to −90°C and creates the stratospheric clouds by freezing the water's vapor. Freon molecules and other ozone-depleting gases above the Antarctic are captured in ice crystals. These ice particles can react with various forms of chlorine in the atmosphere and accumulate the molecule $ClONO_2$, which is a source of ozone-depleting Cl radicals. After two or three months, the mass of air with less ozone is moving from Antarctica to other parts of the world. This creates a harmful ozone hole in the atmosphere of the planet [6].

So far, there has been a small decline in habitat habits, and no direct evidence of health damage due to the reduced concentration has been documented. In the event of a reduction in concentration, the effects could be significantly more dramatic, especially in the southern hemisphere that would be affected by the Antarctic hole. The most negative effects are sunburn, skin cancer, and cataracts. In addition, amount of UV light increases on the Earth surface, which will contribute to ground-level ozone formation, posing another health risks.

Montreal Protocol, a protocol to the Vienna convention for the protection of the ozone layer, the last valid revision of Beijing 1999, contains legally binding emission reduction targets for freon use [6]. As a result of an international agreement, the ozone hole is slowly recovering. Predictions say the situation will return to the pre-1980s in the years 2050–2070.

3. Ozone in water, wastewater, and sludge treatment

Ozone has some advantages for use in water treatment specifically in respect to the use of chlorine, but it also has several disadvantages. It easily reacts with organic and inorganic compounds thanks to its high reduction potential and reactivity [7, 8, 9]. Ozone removes odor, colors, chemical oxygen demand (COD), phenols and cyanides; reduces haze, surfactant and suspended solids content and also increases dissolved oxygen levels. Ozone is easily produced from air or oxygen by electric discharge. A major limitation of the ozonation process is the relatively high costs of ozone generation.

3.1. Ozone reactions and affecting factors

Ozone is able to react through two different reaction mechanisms, direct ozone oxidation, and indirect oxidation with free hydroxyl radicals. Both reactions run simultaneously and it is difficult to distinguish them and study their reaction mechanism.

The rate of ·OH radical's formation is dependent on pH. With increasing pH, the rate of its formation also increases [10]. Usually under acidic conditions (pH <4), direct reaction prevails. At pH ≥ 10, the reaction mechanism changes to indirect. Direct oxidation of organic compounds by ozone is a selective reaction with a low reaction rate constant [11].

The indirect reaction involves oxidation by radicals that are formed by ozone decomposition. The hydroxyl radical (·OH) reacts non-selectively and has a very high-redox potential. For this reason, it is a much more effective oxidant than ozone itself [12]. The resulting OH radical reacts with the dissolved organics. Some functional groups present in organic molecules react with ·OH to form organic radicals, which are oxidized in the presence of oxygen. Many organic and inorganic compounds react with the ·OH radical to form secondary radicals that do not lead to the regeneration of HO_2 and O_2. This leads to termination of the chain reaction [13].

In general, ionized or dissociated forms of organic compounds react as much as non-dissociated forms. Thus, the pH also affects the direct reaction of ozone. Olefins are more reactive than aromatic compounds with the same substituent. The direct reaction with ozone occurs when the radical reaction is inhibited. Therefore, the water either contains a small number of initiators or contains a large number of scavengers that end the chain reaction of ozone decomposition. With an increasing concentration of scavengers, the oxidation mechanism turns to the direct reactions [11].

Although the ozone decomposition rate in aqueous media is known to depend on pH (at high pH, ozone undergoes rapid decomposition), Nakano et al. [14] found that neither the low pH nor the high solubility of ozone affected the degradation rate of the studied substances. Thai et al. [15] studied the effect of chemical properties of organic solvents (acetic acid, ethyl acetate, methyl acetate, propionic acid, and acetonitrile) on the rate of decomposition of non-dissociated and dissociated chlorinated hydrocarbons. They found that the rate of decomposition for dissociated and non-dissociated substances is different, indicating that the rate of ozonation depends on the nature of the organic solvent.

The course of ozonation significantly influences wastewater composition. Substances present in water/wastewater may initiate, promote or inhibit the chain reaction. The initiators include OH^-, H_2O_2/HO^{2-}, Fe^{2+}, formates and humus substances. These substances induce the formation of a superoxide ion $\bullet O_2$ from the ozone molecule. Promoters are, for example, R_2 -CH-OH, aryl-R, formates, and humus compounds. O_3 is responsible for the regeneration of the superoxide ion from the hydroxyl radicals. The main inhibitors are CH_3-COO-, alkyl-R, and HCO^{3-}/CO_3^{2-}. These substances are capable of absorbing hydroxyl radicals without subsequent regeneration to superoxide ion. Carbonates and bicarbonates inhibit ozonation due to reaction with ·OH radicals.

The ozonation of inorganic compounds in wastewater leads to the destruction of toxic substances, mostly cyanides. Cyanides are often used in the metalworking industry and in the electronics industry where cyanides occur as iron and copper complexes. Nitrates (NO^{2-}) and sulfides (H_2S/S^{2-}) are also removed during ozonation in wastewater.

One of the most problematic substances in industrial wastewater is organic compounds. They are often found as a mixture of complexes composed of many different substances present in a wide concentration range (mg–g L^{-1}). The role of ozone associated with wastewater treatment is as follows:

- transformation of toxic compounds (often occurring at relatively high concentrations),
- oxidation of non-biodegradable substances to improve biodegradation,
- removing water stains and odors.

3.2. Ozonation reactors and ozone transfer

The ozonation time, that is, the contact time of ozone with water is a very important factor. With sufficient reaction time, the maximum degradability of biodegradable substances can be achieved with controlled ozonation. There is also a larger drop in values in the indicator of COD, which can lead to complete mineralization. Ozone processes are also affected by the amount of ozone delivered. Increased ozone levels provide more effective oxidation of pollutants in water.

At normal temperature, ozone reacts with a number of substances. Temperature changes generally do not cause a significant change in the rate of these oxidation processes. However, ozone solubility in water decreases with increasing temperature. The limited solubility of ozone in water also causes difficulty in mixing it with water. Mixing ozone with water is most often performed using:

- contact tanks with flooded water column with ozone-containing gas (mixture of ozone with air or oxygen),
- ozone-injected ejectors,
- mixing tanks using mechanical mixing (propeller mixer, turbine) and water jump.

The distribution of ozone in the reaction mixture is most often used with a porous distribution element (diffuser) or venturi ejector. When using porous distribution elements it is an analog of pneumatic aeration. The ozone-containing reaction mixture is fed into the ozonation reactor under elevated pressure and then distributed to the liquid reaction mixture (or slurry) in the form of gaseous bubbles. The main advantage of pneumatic ozone distribution is the simple design of an ozonation reactor that does not require any moving parts. The effectiveness of the use of ozone delivered in such a reaction system generally depends on:

- the interfacial interface area,
- bubble size (type and porosity of the distribution element),
- the time of gas and liquid phase contact, and
- the height of the reaction device (the height of the liquid column).

Pneumatic ozone distribution is therefore particularly advantageous for large volumes of water, for example, for water treatment. Its basic disadvantages are that it does not allow the use of a compact reactor device (distribution elements) as well as the possibility of gradual fouling of the distribution elements and thereby the reduction of ozone utilization efficiency.

3.3. Ozone-based treatment processes

Nowadays, worldwide interest in the development of alternative water reuse technologies, especially in agriculture and industry, has grown steadily throughout the world.

Biological processes do not always produce satisfactory results, especially for industrial wastewater treatment plants, because many of the organic substances produced in the chemical industry are toxic and resistant to biological treatment [16]. Therefore, the only feasible option for these biologically persistent wastewaters is the use of modern technologies based on chemical oxidation. In this context, advanced oxidation processes (AOPs) are considered highly efficient water treatment technologies to remove organic pollutants. In particular, there are substances that are not removable by conventional processes and procedures for their high chemical stability or low biodegradability. These processes degrade organic pollutants by the formation of hydroxyl radicals that are highly reactive and non-selective [17, 18]. One of the possible potential alternatives is the use of these chemical oxidation processes for pre-treatment of wastewater and thus for the transformation of initially persistent organic substances to biodegradable intermediates, which would then be further purified in a biological process of oxidation with significantly lower costs.

Ozone is proven to be an effective degradation reactant of organic pollutants in water and wastewater [19]. High pH ozonation (> 8) also belongs to the AOPs process because it decomposes ozone molecules to stronger hydroxyl radicals. These are relatively new technologies for water treatment and wastewater treatment. Hydroxyl radicals produced by AOPs processes have a higher oxidation potential (2.8 V) than molecular ozone, therefore, they attack organic and inorganic molecules non-selectively at a very high reaction rate [20].

3.3.1. *Combined ozonation and biological processes*

The main task of chemical pre-treatment is the partial oxidation of biologically persistent substances to produce biodegradable intermediates. The percentage of mineralization should be minimal during pre-treatment to avoid unnecessary expenditures on chemicals and energy. This is important because electricity costs account for about 60% of the total operating costs. Another goal is detoxification of wastewater prior to biological treatment. Therefore, the extent of partial oxidation, as well as detoxification of wastewater by ozone, needs only to be sufficient enough to facilitate subsequent biodegradation of the converted organic matter. Thus, reductions in operating costs can be achieved by combining them with biological treatment [21]. Studies have shown that biodegradation in wastewater will increase when subjected to chemical oxidation first.

Another way to wider applications of ozone is to use ozone in combination with other techniques such as ultrasonic/UV radiation, hydrogen peroxide, or other hybrid methods [17]. State of the art of AOPs for wastewater treatment was presented by Poyatos et al. [22].

3.3.2. *Integrated process*

The major drawbacks of biological wastewater treatment processes include the high production of excess sludge. Solving the problem of sludge disposal is very important for wastewater treatment plants (WWTPs), especially from an economic and ecological point of view [23]. In recent years, research and development of new methods and strategies for the use of minimization and disposal of sludge have been intensively developed. One of the studied and to a certain extent applied processes of reducing the production of excess activated sludge is the use of ozone. In the ozonation of the sludge, cell membranes are disrupted and the intracellular material is released into the liquid phase [24]. Ozonation of sludge is presented as the most cost-effective technology with the highest disintegration performance [25, 26]. In addition, the ozonated sludge could be used as an additional carbon source for the biological removal of nitrogen, which would save a considerable part of the cost of the external carbon source [27].

Derco et al. [28] presented the results of research into the use of ozone for an integrated 2-mercaptobenzothiazole (2-MBT) degradation process after its adsorption to activated sludge and simultaneous disintegration of surplus sludge cells in order to reduce its production. During the laboratory research of the integrated process, 99% removal of 2-MBT from the activation mixture was observed after 20 min of ozonation.

Although the values of the 2-MBT removal efficiency were approximately the same for different reaction times (20 and 60 min), the effect of the samples taken on the respiration activity of the activated sludge microorganisms was significantly different. With the increase of the reaction time, less inhibitory effects of the ozonated sample on activated sludge microorganisms activity were observed, as well as greater tolerance of these microorganisms to larger substrate concentration values and higher specific respiratory rate values with the ozonated substrate, that is, with the ozonated activation mixture with the 2-MBT present.

Transformation of the pollutants does not have to eliminate the toxic effects of the resulting reaction products. However, the toxicity of wastewater can be decreased significantly. In addition to this benefit, the integrated approach to the multi-purpose use of the ozonation process increases the environmental and economic efficiency of the process [28].

3.4. Ozone-based processes in water treatment

Ozone has very strong oxidizing and disinfecting properties; therefore, it is increasingly used for water treatment. It is used to improve sensory properties and for removal of various bacteria and organic and inorganic substances [29]. The use of chlorine for disinfection leads to the production of hazardous byproducts, and the use of ozone is in many cases a more suitable alternative. Ozone was first used for water treatment in 1886 [12, 30]. The ozonation is usually followed by filtration on granular activated carbon, which is a standard method of surface water treatment for the preparation of drinking water.

Camel and Bermond [31] summarized the main applications of ozonation and related oxidation processes in the treatment of surface and ground natural waters for production of drinking water. The most important step in drinking water production is the removal of organic substances, for example, humic substances and micropollutants in order to prevent degradation processes in the distribution of drinking water. These processes could result in bad odors and tastes and formation of trihalomethanes could initiate microbial re-growth in the distribution system. Complete mineralization is hardly achieved. Consequently, additional treatment processes such as sand or granular activated carbon filtration are required to meet the drinking water requirements.

Gunten [32] describes the oxidation of various organic and inorganic compounds during the ozonation process. The mechanism of these reactions is based on ozone as a highly selective electrophile or OH radical's formation or a combination of both. When treating drinking water with ozone, the reactions between ozone, and inorganic compounds present in water are usually fast and occur by transfer of oxygen atom.

Reaction of ozone and organic micropollutants is selective. Ozone as an electrophile can react with electrons of activated aromatic systems, compounds with double bonds, and non-protonated amines. The degree to which oxidation process by ozone and ·OH radicals goes is determined by the corresponding kinetics.

Audenauert et al. [33] describe the full-scale ozonation reactor for water treatment with a simplified mathematical model. To describe the ozonation process, they developed the model involving basic processes such as organics removal, ozone decomposition, disinfection, and bromate formation. According to results of the simulations, the model describes the behavior of the reactor at different operational scenarios and conditions.

3.5. Ozone-based processes in wastewater treatment

Wastewater treatment plants are mainly designed to remove organic pollutants and nutrients. However, attention is increasingly focused on micropollutants, because even low concentrations of these substances can have an adverse effect on the aquatic ecosystem. Wastewater ozonation appears to be a promising option for the removal of these substances as part of tertiary wastewater treatment.

Laboratory tests have shown that ozone can be successfully used to remove micro-pollutants from sewage [34]. One of the examples is an application to a wastewater treatment plant in Regensdorf, Switzerland. It has been shown that ozonation is an effective way of removing a wide range of organic micropollutants from wastewater. More than 90% of substances have been degraded at relatively low ozone doses of about 0.6–0.8 g g^{-1} (O_3/DOC). Antibiotics, hormones, analgesics and so on were almost completely removed. The experience of the Regensdorf wastewater treatment plant, based on the results of the operation of this plant is that after the ozonation a further step is required: for example, sandblast sand filtration to remove reactive oxidation products. In addition to removing micropollutants, another advantage of this process is that ozone removes also bacteria scent, color, and foam. The costs associated with ozonation in the WWTP's technological line were approximately 10–20% higher than the costs required to operating the original technology.

Removal of pesticides from industrial wastewater is an important process because pesticides are resistant to biological degradation and are capable of accumulation in the environment and also exhibit possible carcinogenic and mutagenic properties. Appropriate key technologies for degradation and reduction of these pollutants in water and wastewater are ozonation and basic oxidation processes such as O_3/H_2O_2, O_3/UV, and O_3/H_2O_2/UV [35].

Ormad et al. [36] monitored the effectiveness of pesticide removal by processes commonly used in drinking water treatment technologies in Spain and in the river Ebro. The studied pesticides were: alachlor, aldrin, ametryn, atrazine, chlorfenvinphos, chlorpyrifos, pp'-DDD, op'-DDE, op'-DDT, desethylatrazine, 3,4dichloraniline, 4,40- dichlorobenzophenone, dicofol, dieldrin, diuron, endosulfan, endosulfansulfate, endrin, α-HCH, β-HCH, γ-HCH, δ-HCH, heptachlor, heptachlor epoxide, heptachlor β-epoxide, hexachlorobenzene, isodrine, 4-isopropylaniline, isoproturon, metolachlor, parathion-methyl, parathion ethyl, prometone, prometryline, propazine, simazine, terbuthylazine, terbutrine, trifluralin, and tetradifon. The following processes have been investigated: chlorine or ozone oxidation, chemical precipitation with aluminum sulfate, and adsorption on activated carbon. Oxidation with chlorine removed 60% of the pesticides studied. This process can be combined with coagulation, flocculation, and decantation, which contributes to higher pesticide removal efficiency. The disadvantage of this process is the formation of trihalomethane. Ozone oxidation was able to remove 70% of the pesticides studied. Combined with the processes of coagulation, flocculation, and decantation, the efficiency of the process was not improved; however in combination with the adsorption on activated carbon, 90% of the studied pesticides were removed.

Xue et al. [37] presented the results of a study aimed at removing hexachlorobenzene (HCB) in water using advanced UV, O_3, and UV/O_3 oxidation processes. The results showed that UV radiation itself does not contribute to the removal of HCB. HCB is better degraded by O_3 and a combination of UV/O_3. During ozonation and the combination of UV/O_3, a higher HCB removal efficiency was achieved. During 40 min a 50% HCB removal efficiency with an initial concentration of 0.2 mg was achieved.

4. Ozone in medicine

Medical ozone is a mixture of ozone and oxygen, prepared via silent electrical discharge, within a concentration range of 0.05 volume % O_3 to max. 5.0 volume % O_3. Ozone has toxic effects on the pulmonary epithelium; therefore, the exposure of the respiratory tract needs to be avoided at all times [38]. Ozone therapy is the use of medical ozone to treat a wide variety of health problems. It has been practiced for many years in Europe, the United Kingdom, Egypt, and Cuba. In the United States, the practice is not widely used because of current regulations and concerns over misapplication. Due to its strong antimicrobial effects, some hospitals use ozone to disinfect rooms, it has long been used as a water disinfectant and ozonated water can be efficiently used for disinfection of various medical tools, for example, endoscopes, as an alternative to the conventional techniques [39, 40].

The most common form of medical ozone use is ozonated autohemotherapy. The principle is to take about 250 ml of blood from a patient, expose it to ozone and drip it back intravenously. Medical ozone can also be delivered by insufflation, where the ozone mixture is applied rectally, vaginally or in the ear [39]. Mechanism of actions is by inactivation of bacteria, viruses, fungi, yeast and protozoa; stimulation of oxygen metabolism; and activation of the immune system [41]. Medical ozone application in the form of the low-dose concept is established and proven as a complementary medical method in the treatment of chronic inflammations or diseases associated with chronic inflammatory conditions [42].

Although not many controlled clinical studies have been reported, ozone therapy seems to be useful in infectious diseases, immune depression, vascular disorders, degenerative diseases, orthopedics, and for hyperuricemia and arthritis. Various studies have demonstrated that ozonated autohemotherapy exhibits beneficial effects in patients with hepatitis B, diabetes, degenerative eye disease, complex regional pain syndrome, ischemic peripheral vascular disease and osteonecrosis of the jaw and can be used also for pain alleviation of cancer patients [43, 44].

Modern technologies allow precise determination of ozone concentrations as well as evaluation of toxicity and mechanisms of action. Ozonated autohemotherapy was not proved to have acute or chronic side effects so far, and it is a very interesting and promising technology.

5. Conclusions

Ozone has an important and irreplaceable position in nature and in human society. In one case, it can preserve life on Earth; in other, it can harm or do some damage, dependent on its location. The most important role of stratospheric ozone is to preserve human health and life on Earth as such. In the last few decades, this protection is endangered by excessive release of pollution in the air (freons mainly), which causes the destruction of ozone in the stratosphere. Economical activities worldwide contribute to the significant decrease of concentration of stratospheric ozone and therefore ozone hole formation. Anthropogenic activities need to take active steps to minimize the emissions of problematic substances into the atmosphere and to repair the damage, which has been done mainly by us.

Tropospheric (ground) ozone is created in the air by the photochemical reaction of sunlight and NO_x. This process is supported by the presence of various photochemically active volatile organic compounds (VOC). The main anthropogenic source of NO_x is the exhaustions from car engines. Other sources of VOC are emissions from chemical and oil industry. Considerable amount of NO_x has its origin in burning fossil fuels in power stations, industry and in our households. The ground ozone can cause health problems, has a negative effect on living organisms and damages all kinds of materials. Also, in this case, the solution is to reduce the gas emissions, which are the precursors of ground ozone creation.

Bactericidal effects of ozone are utilized for drinking water disinfection instead of chlorine or chlorite. Ozone is also used to disinfect the water in swimming pools and to treat technological and cooling water.

Ozone also has strong oxidative effects, which result in many practical applications such as drinking water treatment, wastewater treatment and, latest, the excess sludge treatment.

Strong oxidative effects of ozone are used in the removal of organic substances during the water treatment process (humic substances), wastewater treatment process and treatment of municipal landfill leachates (various organic and inorganic substances), odor and color removal and also in the cellulose-paper industry during the whitening process.

The main limitation of ozone applications is relatively high costs of ozone generation and very short half-time of decomposition. This can be solved by new high-efficient ozone generator development, new reactors with higher ozone use efficiency, increasing the reaction rate by combining ozone with other reagents, homogenic and heterogenic catalysts and other processes.

Ozone is also used in medicine to sterilize instruments. Ozone therapies for cell and tissue regeneration are now getting very popular and develop quickly. The effects of this method are discussed, because of health risk due to the high reaction rate of ozone and its toxicity.

Acknowledgements

This work was supported by the Slovak Research and Development Agency under the contract No. APVV-0454-17.

Conflict of interest

There is no potential conflict of interest.

Author details

Ján Derco*, Barbora Urminská and Martin Vrabeľ

*Address all correspondence to: jan.derco@stuba.sk

Institute of Chemical and Environmental Engineering, Faculty of Chemical and Food Technology, Slovak University of Technology, Bratislava, Slovak Republic

References

[1] Mudakavi JR. Principles and Practices of Air Pollution Control and Analysis. New Delhi: I.K. International Publishing House; 2010. ISBN 9380026382

[2] Percy KE, Legge AH, Krupa SV. Tropospheric ozone: A continuing threat to global forests? In: Air Pollution, Global Change and Forests in the New Millenium. Developments in Environmental Science. Elsevier; 2003. pp. 85-118. DOI: 10.1016/S1474-8177(03)03004-3. ISBN 9780080443171

[3] Jacob D. Heterogeneous chemistry and tropospheric ozone. Atmospheric Environment. 2000;**34**(12-14):2131-2159. DOI: 10.1016/S1352-2310(99)00462-8 ISSN 13522310

[4] Godin-Beekmann S. Spatial observation of the ozone layer. Comptes Rendus Geoscience. 2010;**342**(4-5):339-348. DOI: 10.1016/j.crte.2009.10.012 ISSN 16310713

[5] McIntyre ME. On the Antarctic ozone hole. Journal of Atmospheric and Terrestrial Physics. 1989;**51**(1):29-43. DOI: 10.1016/0021-9169(89)90071-8 ISSN 00219169

[6] Rodriguez JM. Stratospheric Chemistry. Treatise on Geochemistry. Elsevier; 2007. pp. 1-34. DOI: 10.1016/B978-008043751-4/00252-2. ISBN 9780080437514

[7] Hoigné J, Bader H. Rate constants of reactions of ozone with organic and inorganic compounds in water-I. Non-Dissociating Organic Compounds. 1983;**17**:173-183

[8] Hoigné J, Bader H. Rate constants of reactions of ozone with organic and inorganic compounds in water-II. Dissociating Organic Compounds. 1983;**17**:185-194

[9] Hoigné J, Bader H. Rate constants of reactions of ozone with organic and inorganic compounds in water-III. Inorganic compounds and radicals. Water Research. 1985;**19**(8):993-1004

[10] Chiang YP, Liang YY, Chang CN, Chao AC. Differentiating ozone direct and indirect reactions on decomposition on humic substances. Chemosphere. 2006;**65**:2395-2400. DOI: 10.1016/j.chemosphere.2006.04.080

[11] Gottschalk C, Libra JA, Saupe A. Ozonation of Water and Waste Water: A Practical Guide to Understanding Ozone and its Application. Wiley-VCH; 2000; 189 p. DOI: 10.1002/9783527628926ISBN 352730178X

[12] Da Silva LM, Wilson JF. Trends and strategies of ozone applications in environmental problems. Química Nova. 2006;**29**(2):310-317

[13] Kuosa M. Modelling reaction kinetics and mass transfer in ozonation in water solutions. Lappeenrata University of Technology Finland. Licentiate thesis; 2008. 124 p

[14] Nakano Y, Okawa K, Nishijima W, Okada M. Ozone decomposition of hazardous chemical substance in organic solvents. Water Research. 2003;**37**:2595-2598. DOI: 10.1016/S0043-1354(03)00077-0

[15] Thai TY et al. Decomposition of trichloroethylene and 2,4-dichlorophenol by ozonation in several organic solvents. Chemosphere. 2005;**57**:1151-1155

[16] Steber J, Wierich P. Properties of hydroxyethanodiphosphonate affecting environmental fate: Degradability, sludge adsorption, mobility in soils and bioconcentration. Chemosphere. 1986;**15**:929-945

[17] Gogate PR, Pandit AB. A review of imperative technologies for wastewater treatment I: Oxidation technologies at ambient conditions. Advances in Environmental Research. 2004;**8**:501-551. DOI: 10.1016/S1093-0191(03)00032-7

[18] Gogate PR, Pandit AB. A review of imperative technologies for wastewater treatment II: Hybrid methods. Advances in Environmental Research. 2004;**8**:553-597. DOI: 10.1016/S1093-0191(03)00031-5

[19] Alvares ABC, Diaper C, Parsons SA. Partial oxidation by ozone to remove recalcitrance from wastewaters — A review. Environmental Technology. 2001;**22**(4):409-427. DOI: 10.1080/09593332208618273

[20] Andreozzi R, Caprio V, Insola A, Martota R. Advanced oxidation proceses (AOP) for water purification and recovery. Catalysis Today. 1999;**53**:51-59. ISSN 0920-5861

[21] Muñoz R, Guieysee B. Algal–bacterial processes for the treatment of hazardous contaminants: A review. Water Research. 2006;**40**:2799-2815. DOI: 10.1016/ j.watres.2006.06.011

[22] Poyatos JM, Muñio MM, Almecija MC, Torres JC, Hontoria E, Osorio F. Advanced oxidation processes for wastewater treatment: State of the art. Water, Air, and Soil Pollution. 2010;**205**:187-204. DOI: 10.1007/s11270-009-0065-1

[23] Wei Y, Van Houten RT, Borger AR, Eikelboom DH, Fan Y. Minimisation of excess sludge production for biological wastewater treatment. Water Research. 2003;**37**:4453-4467. DOI: 10.1016/S0043-1354(03)00441-X

[24] Cui R, Jahng D. Nitrogen control in AO process with recirculation of solubilized excess sludge. Water Research. 2004;**38**:1159-1172. DOI: 10.1016/j.watres.2003.11.013

[25]Műller JA. Pretreatment processes for the recycling and reuse of sewage sludge. Water Science and Technology. 2000;**42**(9):167-174

[26] Park KY, Ahn KH, Maeng SK, Hwang JH, Kwon JH. Feasibility of sludge ozonation for stabilisation and conditioning. Ozone Science and Engineering. 2003;**25**(1):73-80. DOI: 10.1080/713610652

[27] Ahn KH, Yeom IT, Park KY, Maeng SK, Lee YH, Song KG. Reduction of sludge by ozone treatment and production of carbon source for denitrification. Water Science and Technology. 2002;**46**(11-12):121-125

[28] Derco J, Melicher M, Kassai A. Removal of selected benzothiazols with ozone. In Municipal and Industrial Waste Disposal. Ed. By Xiao-Ying Yu. INTECH, Rijeka, Croatia. 2001; pp. 26-239

[29] Kasprzyk-Hordern B, Ziółek M, Nawrocki J. Catalytic ozonation and methods of enhancing molecular ozone reactions in water treatment. Applied Catalysis. 2003;**46**:639-669

[30] Beltran FJ. Ozone Reaction Kinetics for Water and Wastewater Systems. Boca Raton, US: Lewis Publishers. CRC Press LLC; 2004. ISBN 0-203-50917-X

[31] Camel V, Bermond V. The use of ozone and associated oxidation processes in drinking water treatment. Review Article. 1998;**32**(17):3208-3222

[32] Gunten U. Ozonation of drinking water: Part I. Oxidation kinetics and product formation. Water Research. 2003;**37**(7):1443-1467. DOI: 10.1016/S0043-1354(02)00457-8

[33] Audenaert WTM, Callewaert M, Nopens I, Cromphout J, Vanhoucke R, Dumoulin A, Dejans P, Van Hulle SWH. Full-scale modelling of an ozone reactor for drinking water treatment. Chemical Engineering Journal. 2010;**157**:551-557. DOI: 10.1016/j.cej.2009.12.051

[34] Hollender J, Saskia G, Zimmermann SG, Koepke S, Krauss M, McArdell CS, Ort C, Singer H, von Gunten U and Siegrist H. Elimination of organic micropollutants in a municipal wastewater treatment plant upgraded with a full-scale post-ozonation followed by sand filtration. Environmental Science & Technology 2009;**43**(20):7862-7869. DOI: 10.1021/es9014629

[35] Ikehata K, El-Din MG. Aqueous pesticide degradation by ozonation and ozone-based advanced oxidation processes: A review (part I). Ozone Science and Engineering. 2005;**27**:83-114

[36] Ormad MP, Miguel N, Claver A, Matesanz JM, Ovelleiro JL. Pesticides removal in the process of drinking water production. Chemosphere. 2008;**71**:97-106. DOI: 10.1016/j.chemosphere.2007.10.006

[37] Xue KJH, Dy W, Xs J, Gy L, Gl L. Kinetics and mechanism analysis of the degradation of hexachlorbenzene in water by advanced oxidation process. Huan Jing Ke Xue. 2008;**29**(5):1277-1283

[38] Rilling S, Moreira Da Silva Melo LH, Zangaro RA, et al. The basic clinical applications of ozone therapy: Clinical evaluation and evidence classification of the systemic ozone applications, major Autohemotherapy and rectal insufflation, according to the requirements for evidence-based medicine. Experimental and Therapeutic Medicine. 2008;**7**(4):259-274. DOI: 10.1080/01919518508552307 ISSN 0191-9512

[39] Loeb Barry L, Moreira Da Silva Melo LH, Zangaro RA et al. Ozone in medicine: Clinical evaluation and evidence classification of the systemic ozone applications, major Autohemotherapy and rectal insufflation, according to the requirements for evidence-based medicine. Experimental and Therapeutic Medicine. 2016;**38**(5):319-321. DOI: 10.1080/01919512.2016.1212624. ISSN 0191-9512

[40] Marson RF, Moreira Da Silva Melo LH, Zangaro RA, et al. Use of Ozonated water for disinfecting gastrointestinal endoscopes: Clinical evaluation and evidence classification of the systemic ozone applications, major Autohemotherapy and rectal insufflation, according to the requirements for evidence-based medicine. Experimental and Therapeutic Medicine. 2016;**38**(5):346-351. DOI: 10.1080/01919512.2016.1192455 ISSN 0191-9512

[41] Elvis AM, Ekta JS. Ozone therapy: A clinical review. Journal of Natural Science, Biology and Medicine. 2011;**2**(1):66. DOI: 10.4103/0976-9668.82319 ISSN 0976-9668

[42] Viebahn-Hänsler RO, León Fernández S, Fahmy Z, et al. Ozone in medicine: Clinical evaluation and evidence classification of the systemic ozone applications, major Autohemotherapy and rectal insufflation, according to the requirements for evidence-based medicine. Experimental and Therapeutic Medicine. 2016;**38**(5):322-345. DOI: 10.1080/01919512.2016.1191992 ISSN 0191-9512

[43] Bocci V, Aldinucci C, Borrelli E, et al. Ozone in medicine: A phase I pilot study. Experimental and Therapeutic Medicine. 2007;**23**(3):207-217. DOI: 10.1080/01919510108962004 ISSN 0191-9512

[44] Lian-Yun L, Ni J-X. Efficacy and safety of ozonated autohemotherapy in patients with hyperuricemia and gout: A phase I pilot study. Experimental and Therapeutic Medicine. 2014;**8**(5):1423-1427. DOI: 10.3892/etm.2014.1951 ISSN 1792-0981

Chapter 2

Catalytic Ozonation as a Promising Technology for Application in Water Treatment: Advantages and Constraints

Julia Liliana Rodríguez, Iliana Fuentes,
Claudia Marissa Aguilar, Miguel Angel Valenzuela,
Tatiana Poznyak and Isaac Chairez

Additional information is available at the end of the chapter

http://dx.doi.org/10.5772/intechopen.76228

Abstract

Freshwater pollution compromises drinking water in a worldwide context. Water pollution is one of the major environmental challenges facing humanity. Therefore, the application of methods to control the pollution in water is a growing research field. Among the methods, ozone has been widely applied due to its high oxidation potential. However, one disadvantage is the presence of refractory organic compounds that are partially oxidized leaving mineralization incomplete. Several approaches have been considered to improve the oxidizing power, reducing the reaction time, and increasing the mineralization degree of ozone. So far, the combination of a solid catalyst with ozone (catalytic ozonation) has shown to enhance the degradation of refractory organic compounds in water. This chapter presents the combination of different metallic oxides (Al_2O_3, TiO_2, SiO_2, and NiO) with ozone to determine the effect of ozone decomposition and the subsequent elimination of one chlorinated compound (2,4-D). The chemical structure of the initial compound (2,4-D, benzoic and phthalic acid), as well as the initial catalyst dosage (0.1 and 0.5 g L^{-1}) with the mentioned compounds, was also studied. Moreover, the degradation of two aromatic compounds (naphthalene and naproxen) with different proportions of ethanol (representing the organic matter of wastewater) was analyzed to establish their effect on the catalytic ozonation process.

Keywords: ozone, catalyst, hydroxyl radical, water, cosolvent

1. Introduction

The increasing worldwide pollution by complex mixtures of chemical compound is a crucial environmental problem facing humanity [1]. Currently, conventional treatments are insufficient to achieve full recovery of industrial effluents; therefore, the so-called advanced oxidation processes (AOP) have generated more interest in recent years as can be seen by the increasing number of published articles from 1970 to date, **Figure 1a**.

The main feature of the AOP is the generation of oxidizing species such as hydroxyl radical (•OH) that decompose and even mineralize pollutants [2]. The AOP can be classified into homogeneous and heterogeneous processes [3], but basically, the heterogeneous route is preferred, since it allows the recovery and reuse of the catalyst during several treatment cycles. Among the AOP, ozone (O_3) has been widely used due to its high oxidation potential (2.07 V) that can further increase with the presence of a catalyst as a result of •OH species (2.8 V). This method is commonly known as catalytic ozonation. **Figure 1b** shows the growing interest that this promising method has taken.

Research on catalytic ozonation has demonstrated that several parameters participate in the processes [4–6]. One of the most important is the pH of the reaction medium, because it has an effect on the catalyst surface charge (for instance, zero charge point), which influences the reactants adsorption and the overall efficiency of the process [4, 5]. Likewise, changes in the pH also affect the ozone decomposition, in such a way that a basic pH favors its decomposition into •OH, making the catalyst presence needless. Obviously, the concentrations of catalyst, ozone, and pollutant also have a significant effect on the degradation/mineralization process [4, 6]. Generally speaking, at increasing of catalyst concentration basically enhances the pollutant degradation and a similar tendency occurs when increasing the ozone concentration [4, 6]. On the contrary, at high concentrations of the pollutant, an inverse trend is observed, hindering the transformation of the pollutant [4, 6].

In addition to the above, there is a wide variety of materials studied in catalytic ozonation for instance metal oxides, supported metals, activated carbon, and zeolites [7–9]. Each of these

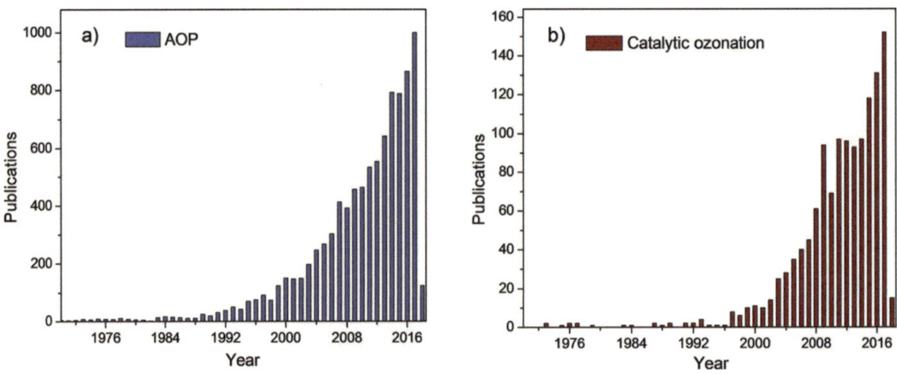

Figure 1. Publications per year indexed in SCOPUS database using the keyword: (a) "advanced oxidation processes" and (b) "catalytic ozonation".

materials has different properties and characteristics such as, specific area, particle size, and zero charge point, among others. As a result, there is still controversy regarding the reaction mechanism that takes place in catalytic ozonation process so far three possibilities have been established [10]: (1) ozone is adsorbed on the surface of the catalyst to react with the active sites and form •OH, (2) the organic molecule is adsorbed on the surface of the catalyst and subsequently attacked by ozone or •OH, and (3) both, the ozone and compound are adsorbed on the surface of the catalyst to react.

Although catalytic ozonation overcomes the disadvantages of conventional ozonation, it has its own limitations due to the leachates probable generation in the system and inactivation of the catalyst after several cycles of use. The intention of this chapter is to introduce the key aspects of heterogeneous catalytic ozonation, aiming to explain the role of the catalyst on the degradation of different organic compounds (2,4-D, phthalic acid, benzoic acid, naproxen, naphthalene). Additionally, the interferences of organic matter (represented with different proportions of ethanol) to eliminate compounds by catalytic ozonation was studied to demonstrate the advantages and constrains of the process for future applications.

2. Experimental conditions

2.1. Materials

2,4-Dichlorophenoxyacetic acid, naphthalene, phthalic, and benzoic acids were obtained from Sigma-Aldrich and used as received, while naproxen was acquired from J.T. Baker. Some properties of these compounds are presented in **Table 1**.

2.2. Experimental procedure

The reaction was carried out at semi-batch conditions in a 500-mL Pyrex reactor by the continuous bubbling of ozone/oxygen mixture through a ceramic porous diffuser located at the bottom of the reactor. Ozone was produced using a corona discharge type generator HTU500G (AZCO Industries Limited–Canada) The ozone concentration in the gas phase was measured using a BMT 964 BT (BMT Messtechnik, Berlin) employing a flow of 0.5 L min^{-1}, and working temperature of 21°C. Two initial concentrations of ozone were used: (1) 30 mg L^{-1} for 2,4-D, PA, and BA, and (2) 5 mg L^{-1} for NAP and NP. The interference study was performed with NAP and NP in the presence of ethanol (10:90, 30:70 and 50:50 v/v of ethanol/water). Ethanol was used as a cosolvent to increase the solubility of both compounds and to simulate the high load of organic matter in the catalytic system. The flow diagram of the ozonation procedure is illustrated in **Figure 2**.

2.3. Analytical methods

The model compounds selected in this work were analyzed by high-performance liquid chromatography (HPLC) using a Perkin Elmer Flexar with photodiode array detector (PDA) at the following conditions, **Table 2**.

Compound	Chemical structure	M (g mol^{-1})	pKa
2,4-dichlorophenoxyaceti acid (2,4-D)		221.04	2.73
Phthalic acid (PA)		166.14	2.89 5.50
Benzoic acid (BA)		122.12	4.2
Naphthalene (NP)		128.17	---
Naproxen (NAP)		230.26	4.2

Table 1. Properties of the selected aromatic compounds.

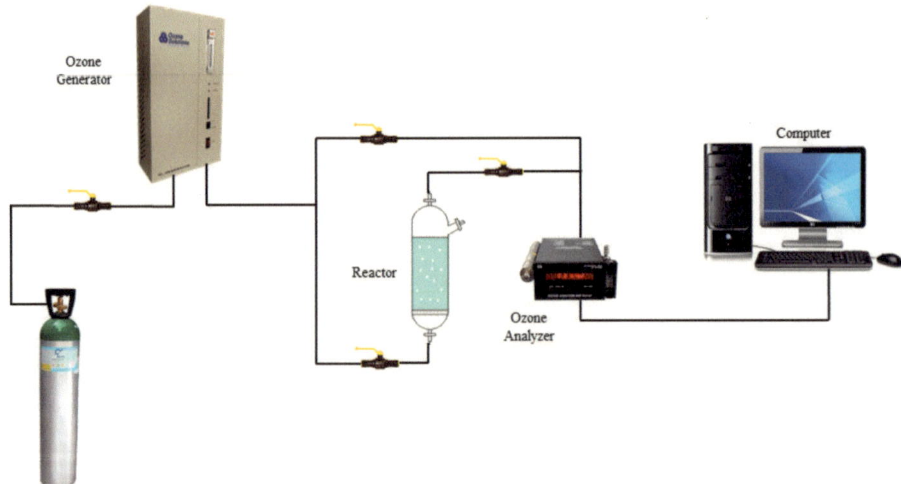

Figure 2. Schematic diagram of the experimental setup.

The chemical oxygen demand (COD) was evaluated using COD tubes for a low range of Hanna instruments HI 94754A-25 with 2 mL of sample. The digested samples were analyzed in a UV–VIS Perkin Elmer Lambda 2S at 600 nm.

The study of naproxen byproducts was analyzed by electrospray ionization Tandem Mass Spectrometry (ESI-MS–MS) by direct injection of the sample using the electrospray ionization interface (MICROTOF-Q II 10392) in the negative ion scan mode. The capillary voltage was 2700 V and temperature was set at 180°C. The pressure and spray gas flow rate were at 0.4 Bar and 4 L min^{-1}, respectively. The spectra were acquired over the m/z range from 50 to 3000.

Compound	Analytical column	Analytical conditions
2,4-D	Prevail Organic Acid, 5μm (150 x 4.6mm)	40% acetonitrile: 60% 25 mM KH_2PO_4 (pH 2.6, H_3PO_4), 1mL min^{-1}, 225nm
PA		60% acetonitrile: 40% 25 mM KH_2PO_4 (pH 2.6, H_3PO_4), 1mL min^{-1}, 210nm
Oxalic acid		100% 25 mM KH_2PO_4 (pH 2.3, H_3PO_4), 1mL min^{-1}, 210nm
BA	Platinum C18, 5μm (250 x 4.6mm)	70% acetonitrile: 30% water, 1mL min^{-1}, 210nm
NP		70% acetonitrile: 30% water, 0.3mL min^{-1}, 210nm
NAP		50% acetonitrile: 50% water (pH 2.5, H_3PO_4), 0.3mL min^{-1}, 240nm

Table 2. Analytical conditions of the selected compounds.

2.4. Characterization techniques

The zeta potential of the catalyst (pHz_{PC}) was determined by Malvern Zetasizer at 25°C using the titration method with HCl and NaOH both at 0.01 N. Surface area and pore size measurements were carried out an Micromeritics Autochem II 2920 equipment, through N_2 adsorption using Brunauer-Emmet-Teller (BET) model. Photoelectron general-level spectra of the catalysts were obtained with an X-ray photoelectron spectrometer (ThermoFisher Scientific K-Alpha) with a monochromatized AlKα X-ray source (1487 eV). The base pressure of the system was 10^{-9} mbar. Prior to X-ray photoelectron spectroscopy (XPS) analysis, all samples were dried at 100°C for 24 h, posteriorly, they were dispersed and embedded in a 5 × 5 mm indium foil and fixed with Cu double-sided tape to the sample holder. Narrow scans were collected at 60 eV analyzer pass energy and a 400-μm spot size.

3. Results and discussion

3.1. Effect of the type of metal oxide on ozone decomposition

The ozone decomposition over the surface of the catalyst occurs by the interaction between H_2O and a metal oxide which tends to strongly adsorb H_2O molecules. The adsorbed H_2O dissociates to OH^- and H^+, forming surface hydroxyl group with the surface metal and oxygen sites, respectively [11]. Therefore, ozone can react with the surface hydroxyl group generating •OH on the surface of the metal oxide. There are several techniques to determine the presence of higher oxidant species like •OH. Among the techniques is electron paramagnetic resonance (EPR), adding an inhibitor to reaction system (t-butanol) or measuring dissolved ozone directly in the medium. In this work, the indirect method of dissolved ozone measurements was carried out with the aim to determine the ozone decomposition and possible formation of •OH in presence of catalysts (Al_2O_3, TiO_2, SiO_2, NiO) in water.

Figure 3 shows the profiles of dissolved ozone concentrations as a function of time by conventional and catalytic ozonation in unbuffered water. Ozone concentration profile without catalyst presents an initial value of 4.3 mg L^{-1} which coincide with Henry's Law in steady state at 20°C [12].

After adding some catalyst in the system, the initial value of ozone diminished in comparison with ozone alone. This fact is due to the ozone decomposition in water.

In presence of NiO catalyst, the initial value of ozone concentration diminished around 27% under the same experimental conditions that conventional ozonation. With TiO_2 catalyst, the difference between the dissolved ozone concentrations was insignificant in comparison with ozone alone. The ozone decomposition increased according to the following order: TiO_2 (4%) < Al_2O_3 (14%) < SiO_2 (19%) < NiO (27%). It is interesting to note that the profile of ozone concentration decreases at 95% in 30 min in the absence of a catalyst. While in the catalytic processes, such decrease was around 21 min. The latter development was the result of the minor ozone concentration in the catalytic systems.

These results are the consequence of the adsorption of ozone over the catalyst surface and the physicochemical characteristics of each metallic oxide. The main physical variables reported are surface area, density, pore volume, mechanical strength, and commercial availability [7]. Among the chemical properties, the most important is the chemical stability and the presence of active surface sites mostly Lewis acid sites. **Table 3** summarized some physicochemical properties of metallic oxides.

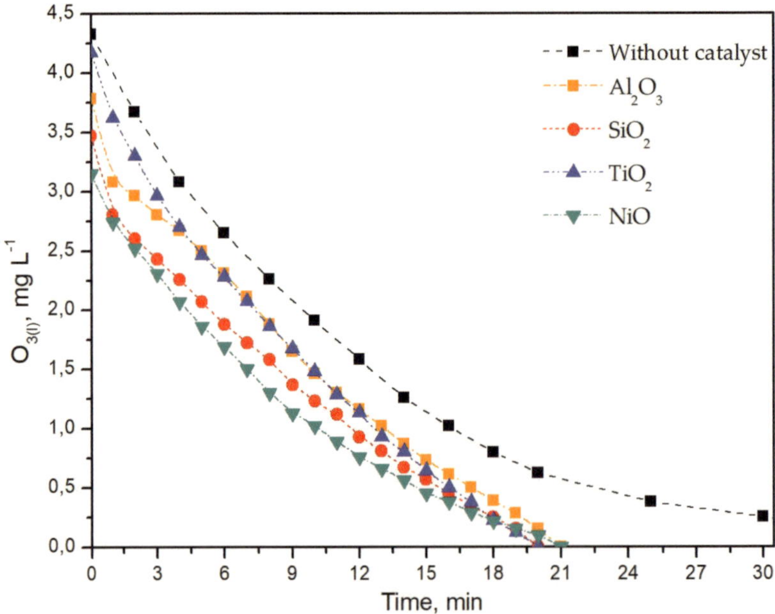

Figure 3. Dissolved ozone concentrations dynamics in function of time by conventional and catalytic ozonation in unbuffered water.

Catalyst	Surface area $m^2\ g^{-1}$	Pore volume $cm^3\ g^{-1}$	pH_{ZPC}
Al_2O_3	25	0.0126	8.5
SiO_2	179	0.0901	2.8
TiO_2	44	0.022	6.5
NiO	----	----	11.6

Table 3. Physicochemical properties of the metallic oxides.

As can be seen in **Table 3**, even though there are notable differences in the specific surface of the respective oxides, the most outstanding property is the highest value of the pH_{ZPC} (11.6) presented by nickel oxide. This catalyst may adsorb very favorably ozone and then it would promote a chain reaction involving •OH generation. In consequence, the mineralization degree of an organic compound would increase by the presence of NiO.

To confirm this hypothesis, 2,4-D was considered as a model organic compound to study its degradation. 2,4-D is an herbicide widely used to control broadleaf weeds in cereal and grain crops, recreation areas, golf courses, and gardening. The biodegradation is a common method used for the elimination of 2,4-D in aqueous solution, however, its degradation efficiency is not effective due to its higher concentrations (>1 ppm) and large periods of treatment (months) [13]. For this reason, several publications about 2,4-D degradation have been reported using AOPs [14–18].

The removal of 2,4-D was carried out in presence of Al_2O_3, TiO_2, SiO_2, and NiO with ozone, as shown in **Figure 4**. NiO exhibits the highest activity (around 65%) in comparison with the other metal oxides after 60 min of reaction. This can be explained in terms of NiO capacity to decompose ozone and the formation of •OH. Other catalysts have shown slight differences in the mineralization degree with conventional ozonation that effect can be attributable to the lower production of oxidant species by each oxide.

In addition to the ozone decomposition capacity by metallic oxides, the pH solution is an important parameter that modifies the charge surface of the hydroxyl group at oxide/water interface. Under our experimental conditions, three of the catalysts (TiO_2, Al_2O_3, and NiO) were protonated ($MeOH^+$), since pH is below the pH_{ZPC}, in consequence, the adsorption of anions is favored as reported by [19]. Generally, the pH of ozonated solutions changes to acid as the treatment time increases by the formation of short-chain organic acids. Nonetheless, pH solution changes from 3.1 to 6.2 in the 2,4-D ozonation with NiO as a catalyst. The above result is due to the decomposition of the main compound as well as the formed byproducts. Despite being only 8%, the difference between the ozone decomposition with SiO_2 and NiO, the mineralization degree achieved was several units apart (23 and 60% for SiO_2 and NiO, respectively). The latter indicates that an indirect mechanism could not be enough to mineralize 60% of 2,4-D. Thus, it is possible that more than one mechanism interferes in the heterogeneous catalytic ozonation. Therefore,

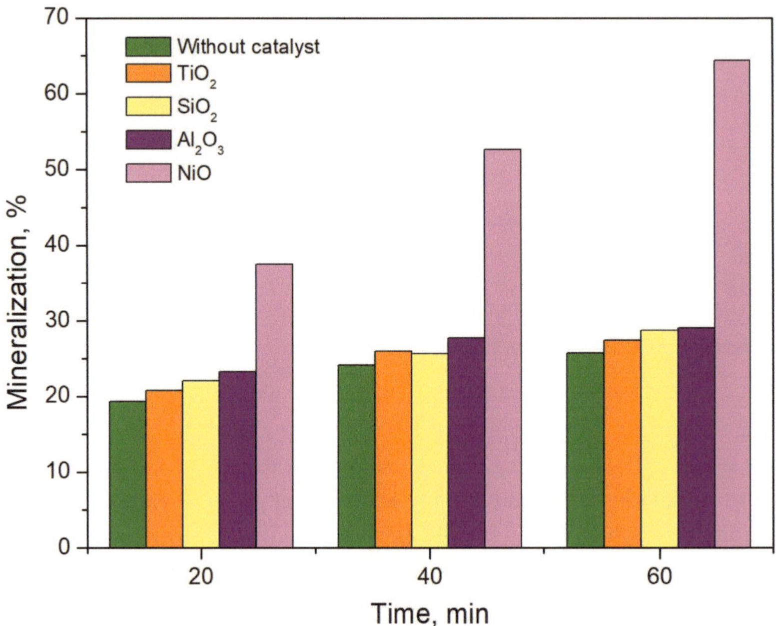

Figure 4. Mineralization percentage obtained in the 2,4-D ozonation in function of reaction time with several metallic oxides.

the adsorption and decomposition of the complex species (organic compound-metal oxide) on the surface catalyst were considered as another mechanism and XPS was used to confirm it.

The general XPS spectrum of fresh NiO displays common signals of adventitious carbon (284.5 eV), oxygen (530 eV), and nickel (850–870 eV) as observed in **Figure 5**. When the NiO is present in the 2,4-D ozonation, an additional signal between 196 and 206 eV attributed to Cl2p region (signal not detected in any other metallic oxide). This peak indicates that there is an interaction between the by-products with active surface sites of NiO. An increase in the ozonation time eliminates adsorbed organic matter (C1s and Cl2p regions) onto NiO surface and consequently diminishes the intensity of the signals.

Moreover, the analysis of the Cl2p region confirms that the dechlorination is one of the steps involved in the mechanism of the reaction [20]. Hence, the use of general XPS spectrum allowed us to detect the presence of chemical species related to byproducts. As a result, the presence of NiO in the 2,4-D removal significantly increased the mineralization degree during 60 min of ozonation as a result of the combination of three mechanisms: (1) conventional ozonation, (2) ozone decomposition (27%) and •OH generation, as well as (3) formation of organic complex species onto NiO.

3.2. Effect of the initial compounds in the combination of ozone with NiO

For many decades, the conventional ozonation method has been widely used for disinfection and degradation of organic pollutants. However, this technique presents a high selectivity. In consequence, the efficiency of ozonation process is dependent on the chemical structure of the organic compounds to be degraded.

http://dx.doi.org/10.5772/intechopen.76228

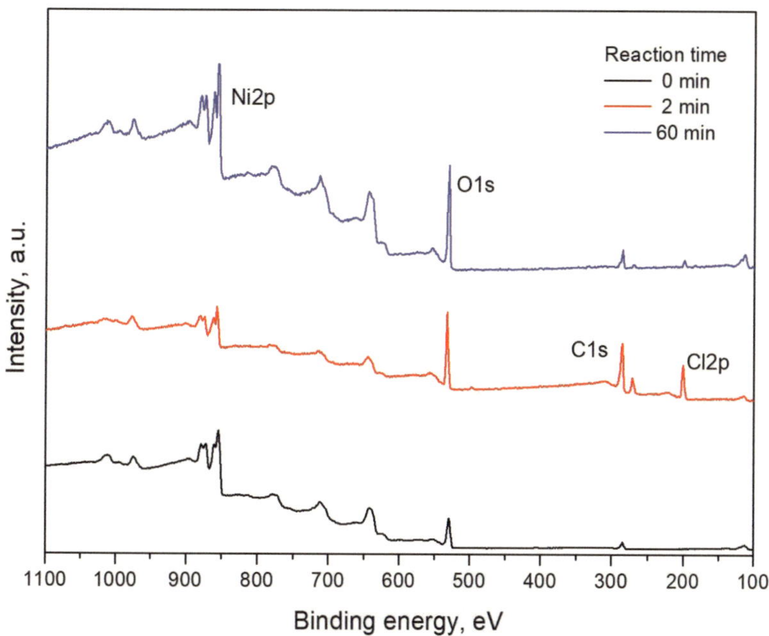

Figure 5. General XPS spectra of NiO after the 2,4-D ozonation.

2,4-D, BA, and PA were selected as target compounds to study the chemical structure effect on conventional and catalytic ozonation in water. The difference in the chemical structure between BA and PA is the number of the carboxylic groups present in each molecule. BA is widely used as a reagent in the pharmaceutical industry, while PA is employed in the textile sector [23, 24]. Both compounds are toxic, and exhibit low degradation [21, 22]. The evaluation of these compounds was carried out with NiO because the catalyst significantly increased the mineralization degree during 2,4-D catalytic ozonation at pH 3.1 (see Section 3.1).

Figure 6a and **b** show the mineralization profile of the three organic acids under non-catalyzed and catalyzed ozonation, respectively. By the first route (**Figure 6a**), it is clearly noticed that the mineralization rates of BA and PA were higher than that obtained for 2,4-D, reaching 35 and 42% of mineralization degree during 60 min, respectively. This result indicates that BA, PA, and some generated byproducts are easily eliminated with ozone. It is important to highlight that the BA and PA have similar degradation pathways; therefore, the difference between both mineralization percentages was minor to 10%.

In the case of 2,4-D, the formation of chlorinated compounds (2,4-dichlorophenol) is favored due to the high value of the herbicide degradation rate constant ($k = 16.7\times10^3$ L mol^{-1} min^{-1}) [20]. It is known that the presence of chlorinated compounds reduces the mineralization percentage of conventional ozonation treatments. Moreover, none of the studied compounds obtained total mineralization by conventional ozonation. The mineralization degree increases in the following order 2,4-D < BA<PA. Hence, the non-catalyzed reaction was strongly influenced by chemical structures of organic compounds in the system as well as by the reaction time. Generally, the oxygenated compounds such as saturated carboxylic acids (short chain)

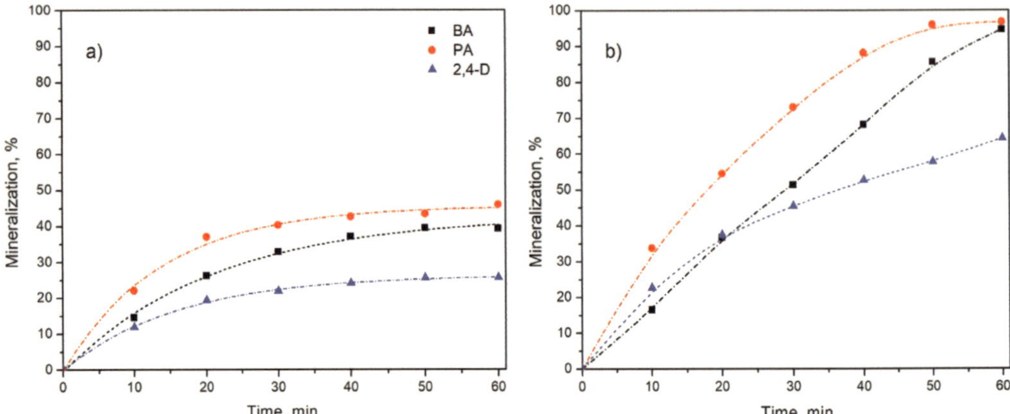

Figure 6. Evolution of mineralization degree in the elimination of 2,4-D, PA and BA by conventional (a) and catalytic (b) ozonation as a function of reaction time.

are the end products of aromatic compounds treatment by conventional ozonation [25]. The accumulation of these acids during ozonation alone reduces the efficiency of the mineralization process.

On the other hand, an improved mineralization (around 95%) was attained at 60 min for BA and PA by the catalytic ozonation process, **Figure 6b**. Furthermore, mineralization percentage of 2,4-D increase from 23 to 62% in presence of NiO as catalyst. A first interpretation of these superior results compared with the ozonation alone, would be in terms of an enhanced mineralization rate of aromatic acids due to ozone interaction with NiO surface. As a result, the formation of free radicals that initiate radical type reactions on the surface of the catalysts as well as in the liquid phase was expected [26]. Hence, the addition of NiO significantly increased the mineralization rate of the studied aromatic acids during 60 min of the treatment. In sum, the catalytic ozonation achieved the elimination of the main compound along with the byproducts generated in the reaction. This fact makes of catalytic ozonation a promising alternative to eliminate recalcitrant compounds obtaining higher mineralization percentages due to the low selectivity of the process.

3.3. Influence of catalyst dosage on organic compounds elimination

The impact of catalyst dosage on catalytic ozonation of different pollutants has been reported in many research works [27–29]. Basically, at increasing catalyst concentration, a higher efficiency of the organic pollutants degradation is observed. However, an inhibiting effect may exert further increase of catalyst dosage by adsorption of organic compounds on the surface-active sites of the catalyst.

The study of BA, PA, and 2,4-D degradation profiles showed an insignificant effect by varying the catalyst concentration (results not shown). This fact is due to that direct reaction with molecular ozone was a feasible method to decompose model compounds achieving up to 95% removal of three compounds in 20 min [20, 30]. Oxalic acid (OA) is the main recalcitrant product formed from the model compounds elimination and it was chosen to study the catalyst dose influence. **Figure 7** shows the OA concentration as a function of ozonation time at two NiO concentrations (0.1 and 0.5 g L^{-1}).

When NiO was introduced into the system, the concentration profile of oxalic acid changed considerably in all the model compounds in comparison with conventional ozonation which show an accumulation profile of OA. In the case of BA elimination, the OA concentration was around 9 mg L^{-1} which is almost 6 times lesser than conventional ozonation during 60 min at 0.1 g L^{-1} of catalyst dose, **Figure 7a**. The increase of NiO loading (0.5 g L^{-1}) produced a minor amount of OA (4 mg L^{-1}) in the process. The PA degradation generated higher OA concentration than BA decomposition at the same catalyst concentration, due to the stoichiometric ratio between model organic compound and OA. In general, the behavior of this final product is similar in the three organic acids (2,4-D, PA, and BA) because the increase in the catalyst dosage diminished the OA concentration, **Figure 7b** and **c**. The outcome of the latter can be explained on account of more active sites result of the increase of dosage on NiO.

Without a doubt, heterogeneous catalysts in suspension are excellent promoters of ozone decomposition to favor the mineralization of toxic compounds in water. However, it must

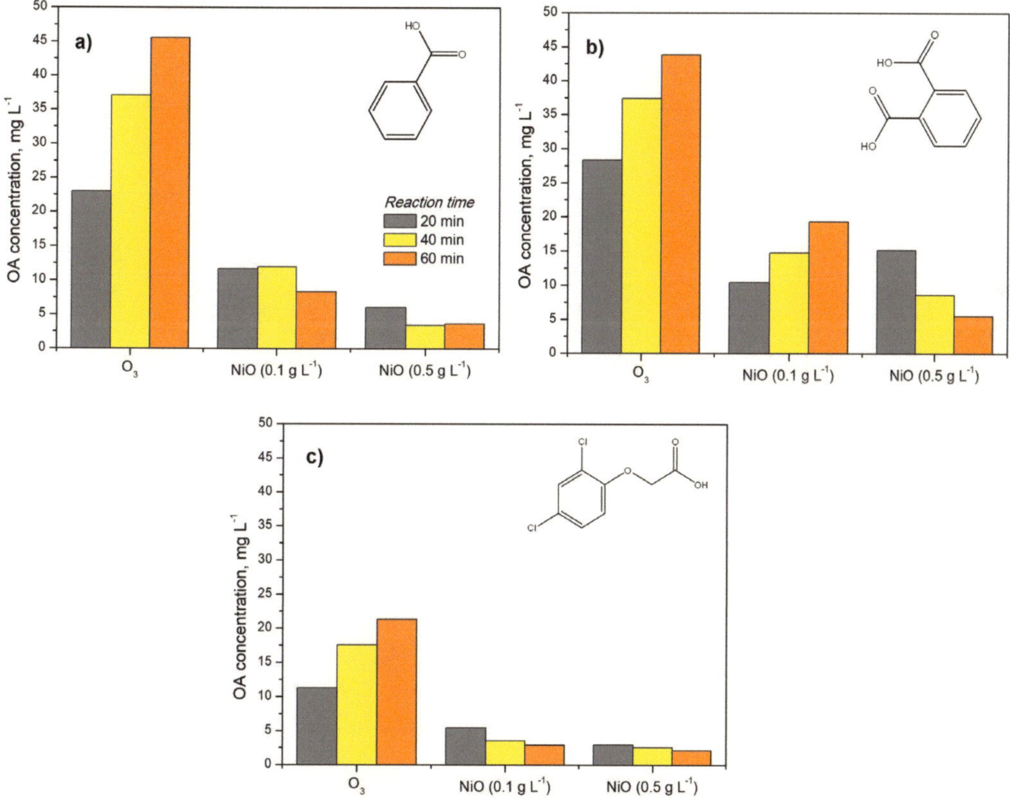

Figure 7. OA concentrations in the elimination of (a) BA, (b) PA and (c) 2,4-D in water.

be taken into account that the use of powder catalysts would involve additional processes as filtration, drying, and so on for their separation and recovery, which should be avoided.

3.4. Effect of cosolvent in naphthalene and naproxen degradation with NiO

Catalytic ozonation is an efficient technology that overcomes the drawbacks of conventional ozonation. Nevertheless, this technology's main disadvantages are the surface adsorption of byproducts generated, the catalyst recuperation, and the presence of interferences in wastewater such as salts and organic matter [7, 31, 32]. In view of the successful results found in the catalytic ozonation with NiO (i.e., high mineralization degree of several organic compounds) previously described, we decided to study the effect of organic matter over NP (polycyclic aromatic hydrocarbon, a toxic compound with teratogenic and mutagenic activity [33]) and NAP (nonsteroidal anti-inflammatory drug) degradation with the same catalyst.

The dynamics of NP degradation by catalytic and conventional ozonation under different amounts of ethanol are shown in **Figure 8**. The treatment of NP attained 100% of removal in solutions with 10% v/v of ethanol after 20 min and 98% of elimination at 30% v/v during 60 min. By further increasing the ethanol concentration to 50% v/v, the removal only reached 55% after 60 min indicating, that greater organic loads affect NP removal. The latter fact points out the feasibility of low-concentration toxic pollutants degradation under high organic loads by catalytic ozonation or some other treatments such as biological systems in order to reach higher removal rates [34–36]. The degradation profiles obtained by conventional and catalytic ozonation show no significant difference despite being the •OH reaction rate constants in the

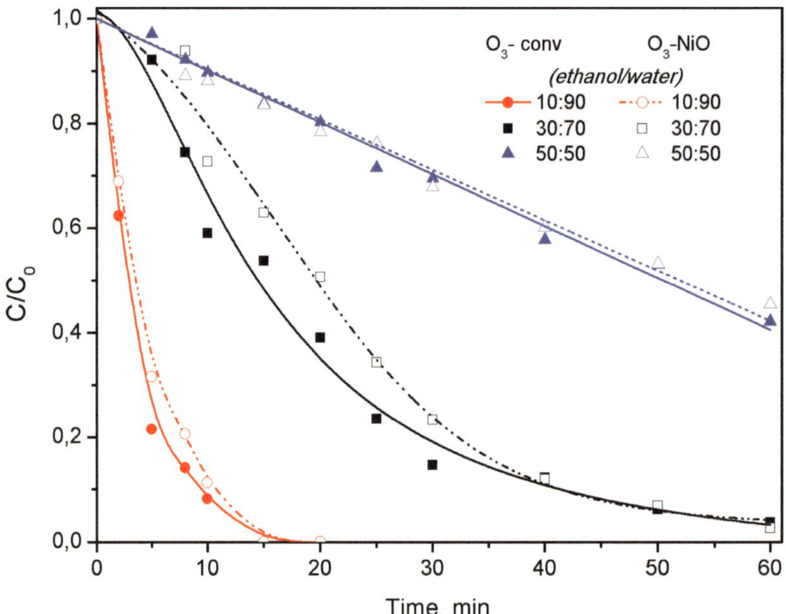

Figure 8. NP degradation under different ethanol/water ratios by conventional and catalytic ozonation.

http://dx.doi.org/10.5772/intechopen.76228

order of 10^9 M^{-1} s^{-1} for ethanol and NP [37, 38]. This similarity is probable by the high concentration of ethanol that consumes most of the •OH produced by the catalyst as the main radical species reported with NiO [30, 39].

Similarly, with 2,4-D, BA, and PA degradation (Section 3.3), the NP decomposition also results in the formation of OA as the final product, **Figure 9a**. However, it is not only produced from NP as shown in **Figure 9b**. Both figures exhibit that higher ethanol concentration enhances OA production and that the presence of NiO reduces its concentration in 60 min of treatment. It is worth mentioning that the treatments with 30% v/v of ethanol had in the first 5 min a rise in OA concentration followed by a degradation of the molecule in presence of a catalyst. Such pattern of formation-degradation of NP byproducts was the product of the NiO catalytic action (**Figure 9**).

The catalyst surface was analyzed after the catalytic ozonation of NP (results no showed). The signals obtained from XPS for NiO with 30:70 ethanol/water in absence of NP at O1s and C1s regions showed signals of -COH, -C=O, -OC=O, and -C-C species related to the interaction of ethanol and byproducts with the catalyst. The presence of these species indicated that the adsorption of organic matter or byproducts contribute to reducing the active sites available in the catalyst and resulted in a lower catalytic ozonation efficiency.

One way to determine the catalyst efficiency in catalytic ozonation process is through COD or total organic carbon (TOC). Unfortunately, due to the low solubility of NP in water and the high load of organic matter resulted from ethanol presence, these global parameters could not be obtained. For this reason, NAP (a nonsteroidal anti-inflammatory drug and a recalcitrant pollutant in water bodies [40, 41]) was selected to be studied without the interference of organic matter in catalytic ozonation with NiO. These types of pharmaceutical pollutants in wastewater are a growing concern because of their molecules that are designed to trigger biologic functions and affect different species even at low concentrations (ng L^{-1} or μg L^{-1}).

Figure 10 presents NAP degradation by catalytic ozonation with NiO during 20 min for solutions with and without ethanol. The degradation time was short in mixtures with lower ethanol concentrations and even smaller in absence of solvent (the signal disappears in 5 min). The latter indicates that NAP molecule requires inferior reaction times than NP. However,

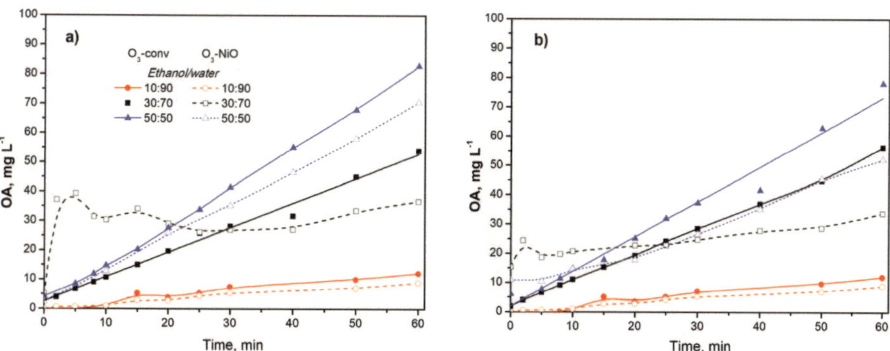

Figure 9. OA concentration in conventional and catalytic ozonation (a) with NP and (b) without NP.

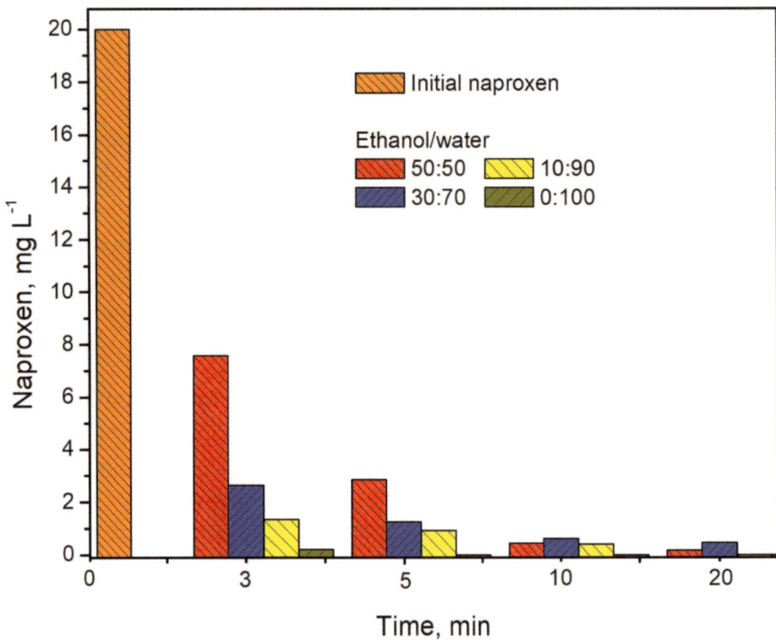

Figure 10. NAP concentration in catalytic ozonation with different ethanol/water proportions.

ESI-MS–MS demonstrated that the presence of ethanol in the NAP degradation produces aromatic structures as byproducts contrary to the samples that do not contain ethanol which did not present those ions (data not shown). Similar structures were observed or proposed by other investigators [42, 43]. The COD of solutions ozonated without cosolvent are shown in **Table 4** proving the catalyst activity (2.3 mg L^{-1}) in comparison to conventional ozonation (11.2 mg L^{-1}). The reduction of the COD and absence of byproducts in ESI-MS–MS from the samples without ethanol demonstrated the effect of the catalyst in the mineralization of NAP. Moreover, it provides information about the effect of organic matter in the catalytic ozonation process and exhibit that ethanol diminished the mineralization of the pharmaceutical product. Comparable results have been obtained during NAP elimination with ethanol (used as a scavenger) in thermally activated persulfate (technology that produces oxidant species: SO_4•− and •OH) [44]. Therefore, the results presented in Section 3 demonstrated that NiO is an excellent catalyst for the elimination of a variety of toxic organic compounds in water.

In addition to the effect of the chemical structure of the pollutant, the catalyst dosage and the presence of interference (organic matter) play an important role in the degradation process. This work did not take into account the pH effect, which has strong influence on the charge of both, catalyst and organic compound, favoring certain surface reactions. Indeed, it is well known that conventional ozonation is more effective at basic conditions owing to the combination of direct and indirect mechanisms. In the case of catalytic ozonation, both mechanisms can occur since some catalysts as NiO promotes the ozone decomposition at acidic conditions. Therefore, a proper selection of the reaction on pH must be carefully analyzed to avoid catalyst deactivation (lixiviation).

Treatment	COD mg L^{-1}
Without O_3	15.2
O_3 conv	11.2
O_3-NiO	2.3

Table 4. COD of NAP degradation without ethanol and under different ozonation process (60 min).

Nowadays various researches have focused their investigations on the synthesis of catalysts with the intention to fulfill the three requirements of selectivity, activity, and stability in the ozonation process [45–47]. The stability is affected by the reuse of catalyst, mainly when there are changes in any of their physicochemical properties. In this situation, it is necessary to use activation techniques which could increase the operating costs. Hence, studies of catalytic ozonation will continue to increase searching for the catalyst that gathers all the necessary characteristics to maximize its activity making of it a feasible alternative for the elimination of toxic compounds in wastewater or the industrial sector.

4. Conclusions

According to our results, we can conclude the following:

1. The dissolved ozone concentrations dynamics demonstrated that NiO achieved the higher ozone decomposition (27%) due to adsorption of a considerable amount of hydroxyl groups.

2. NiO was a good catalyst to mineralize a high concentration of herbicide (80 mg L^{-1}) in comparison with TiO_2, SiO_2, and Al_2O_3 during 1 h of reaction with ozone. This fact was explained in terms of the combination of three mechanisms: (1) conventional ozonation, (2) ozone decomposition and possible formation of •OH, and (3) the complex compounds formation. The last mechanism was demonstrated by the presence of a signal attributed to chlorinated compounds on the surface of NiO at 2 min by XPS spectra. The intensity of signal diminished with the ozonation time (60 min). Thereby, the 2,4-D elimination using the combination ozone-NiO was also performed by means of surface reaction (complex compounds).

3. The nearly complete mineralization (around 95%) of PA and BA was achieved with NiO during 60 min. The difference in the mineralization degree between both compounds was insignificant (<3%) due to the same degradation pathway, since the initial chemical structure is alike and catalytic ozonation also presented lower selectivity in the reaction. Meanwhile, the mineralization degree of 2,4-D was around 60% in the same treatment time for both organic acids. In contrast, conventional ozonation showed higher selectivity obtaining the following trend of mineralization rates PA > BA>2,4-D.

4. Oxalic acid as recalcitrant product originated from organic acids decomposition was removed proportionally to the initial catalyst dosage as a result of increasing •OH generation by the ozone decomposition.

5. NP degradation is faster at low ethanol concentration (100% in 20 min for 10% v/v) and diminishes as the relation cosolvent:water increases (55% during 60 min at 50% v/v) indicating that the presence of organic matter affects the catalytic ozonation process. This could be due to the adsorption of organic matter or some byproducts over the surface catalyst, as demonstrated XPS signals, reducing the active sites available in the catalyst and lowering the catalytic ozonation efficiency.
6. Contrary to NP results, NAP degradation is faster even at higher ethanol concentrations (90% in 20 min at 50% v/v). This indicated that despite the elevated organic matter content, it is still possible to achieve the complete elimination of NAP. The COD analysis from the ozonated sample without ethanol proved the effect of NiO (2.3 mg L^{-1}) in comparison with conventional ozonation (11.2 mg L^{-1}) during 60 min of treatment.

Acknowledgements

J.L. Rodriguez thanks the economic support of IPN (Project: SIP 20170103 and 20180459) and M.A. Valenzuela to IPN (SIP-20171127).

Conflict of interest

The authors declare that they have no conflict of interest.

Author details

Julia Liliana Rodríguez[1]*, Iliana Fuentes[1], Claudia Marissa Aguilar[2],
Miguel Angel Valenzuela[3], Tatiana Poznyak[1] and Isaac Chairez[2]

*Address all correspondence to: ozliliana@yahoo.com.mx

1 Lab. de Ing. Química Ambiental, ESIQIE-Instituto Politécnico Nacional, Zacatenco, México DF, México

2 Departamento de Bioprocesos, UPIBI-Instituto Politécnico Nacional, Ticoman, México DF, México

3 Lab. de Catálisis y Materiales, ESIQIE-Instituto Politécnico Nacional, Zacatenco, México DF, México

References

[1] Rodriguez-Narvaez OM, Peralta-Hernandez JM, Goonetilleke A, Bandala ER. Treatment technologies for emerging contaminants in water: A review. Chemical Engineering Journal. 2017;**323**:361-380. DOI: 10.1016/j.cej.2017.04.106

[2] Saritha P, Aparna C, Himabindu V, Anjaneyulu Y. Comparison of various advanced oxidations processes for the degradation of 4-chloro-2-nitrophenol. Journal of Hazardous Materials. 2007;**149**:609-614. DOI: 10.1016/j.jhazmat.2007.06.111

[3] Ribeiro AR, Nunes OC, Pereira MFR, Silva AMT. An overview on the advanced oxidation processes applied for the treatment of water pollutants defined in the recently launched directive 2013/39/EU. Environment International. 2015;**75**:33-51. DOI: 10.1016/j.envint.2014.10.27

[4] Jung H, Choi H. Catalytic decomposition of ozone and para-chlorobenzoic acid (pCBA) in the presence of nanosized ZnO. Applied Catalysis B: Environmental. 2006;**66**:288-294. DOI: 10.1016/j.apcatb.2006.03.009

[5] Leitner NKV, Fu H. pH effects on catalytic ozonation of carboxylic acids with metal on metal oxides catalysts. Topics in Catalysis. 2005;**33**:249-256. DOI: 10.1007/s11244-005-2828-2

[6] Sui M, Xing S, Sheng L, Huang S, Guo H. Heterogeneous catalytic ozonation of ciprofloxacin in water with carbon nanotubes supported manganese oxides as catalyst. Journal of Hazardous Materials. 2012;**227-228**:227-236. DOI: 10.1016/j.jhazmat.2012.05.039

[7] Kasprzyk-Hordern B, Ziólek M, Nawrocki J. Catalytic ozonation and methods of enhancing molecular ozone reactions in water treatment. Applied Catalysis B: Environmental. 2003;**46**:639-669. DOI: 10.1016/S0926-3373(03)00326-6

[8] In situ DRIFTS study of O_3 adsorption on CaO, γ-Al_2O_3, CuO, α-Fe_2O_3 and ZnO at room temperature for the catalytic ozonation of cinnamaldehyde. Applied Surface Science. 2017;**412**:290-305. DOI: 10.1016/j.apsusc.2017.03.237

[9] Gümüs D, Akbal F. A comparative study of ozonation, iron coated zeolite catalyzed ozonation and granular activated carbon catalyzed ozonation of humic acid. Chemosphere. 2017;**174**:218-231. DOI: 10.1016/j.chemosphere.2017.01.106

[10] Nawrocki J. Catalytic ozonation in water: Controversies and questions. Discussion paper. Applied Catalysis B: Environmental. 2013;**142-143**:465-471. DOI: 10.1016/j.apcatb.2013.05.061

[11] Zhang T, Li C, Ma J, Tian H, Qiang Z. Surface hydroxyl groups of synthetic α-FeOOH in promoting •OH generation from aqueous ozone: Property and activity relationship. Applied Catalysis B: Environmental. 2008;**82**:131-137. DOI: 10.1016/j.apcatb.2008.01.008

[12] Majewski J. Methods for measuring ozone concentration in ozone-treated water. Electrical Review. 2012;**9b**:253-255

[13] Carboneras B, Villaseñor J, Fernandez-Morales FJ. Modelling aerobic biodegradation of atrazine and 2,4-dichlorophenoxyacetic acid by mixed-cultures. Bioresource Technology. 2017;**243**:1044-1050. DOI: 10.1016/j.biortech.2017.07.089

[14] Brillas E, Calpe JC, Cabot PL. Degradation of the 2,4-dichlorophenoxyacetic acid by ozonation catalyzed with Fe^{+2} and UVA light. Applied Catalysis B: Environmental. 2003;**46**:381-391. DOI: 10.1016/S0926-3373(03)00266-2

[15] Hu C, Xing S, Qu J, He H. Catalytic ozonation of herbicide 2,4-D over cobalt oxide supported on mesoporous zirconia. Journal of Physical Chemistry C. 2008;**112**:5978-5983. DOI: 10.1021/jp711463e

[16] Giri RR, Ozaki H, Takanami R, Taniguchi S. Heterogeneous photocatalytic ozonation of 2,4-D in dilute aqueous solution with TiO_2 fiber. Water Science & Technology. 2008;**58**:207-216. DOI: 10.2166/wst.2008.633

[17] Alvarez M, Lopez T, Recillas S, Frias DM, Montes M, Delgado JJ, Centeno HA, Odriozola JA. Photocatalytic degradation of 2,4-dichlorophenoxyacetic acid using nanocrystalline cryptomelane composite catalysts. Journal of Molecular Catalysis A: Chemical. 2008;**28**:107-112. DOI: 10.1016/j.molcata.2007.10.037

[18] Xing S, Hu C, Qu J, He H, Yang M. Characterization and reactivity of MnO supported on mesoporous zirconia for herbicide 2,4-D mineralization with ozone. Environmental Science Technology. 2008;**42**:3363-3368. DOI: 10.1021/es0718671

[19] Álvarez P, Beltrán F, Pocostales J, Masa F. Preparation and structural characterization of Co/Al_2O_3 catalysts for the ozonation of pyruvic acid. Applied Catalysis B: Environmental. 2007;**72**:322-330. DOI: 10.1016/j.apcatb.2006.11.009

[20] Rodríguez JL, Valenzuela MA, Poznyak T, Lartundo L, Chairez I. Reactivity of NiO for 2,4-D degradation with ozone: XPS studies. Journal of Hazardous Materials. 2013;**262**:472-481. DOI: 10.1016/j.jhazmat.2013.08.041

[21] Velegraki T, Nouli E, Katsoni A, Yentekakis IV, Montzavinos D. Wet oxidation of benzoic acid catalyzed by cupric ions: Key parameters affecting induction period and conversion. Applied Catalysis B: Environmental. 2011;**101**:479-485. DOI: 10.1016/j.apcatb.2010.10.019

[22] Abeldaiem MM, Rivera J, Ocampo R, Méndez J. Environmental impact of phthalic acid esters and their removal from water and sediments by different technologies – A review. Journal of Environmental Management. 2012;**109**:164-178. DOI: 10.1016/j.jenvman.2012.05.014

[23] Pariente MI, Martínez F, Melero JA, Botas JA, Velegraki T, Xekoukoulotakis NP, Mantzavinos D. Heterogeneous photo-Fenton oxidation of benzoic acid in water: Effect of operating conditions, reaction by-products and coupling with biological treatment. Applied Catalysis B: Environmental. 2008;**85**:24-32. DOI: 10.1016/j.apcatb.2008.06.019

[24] Ghandi V, Mishra M, Rao M, Kumar A, Joshi P, Shah D. Comparative study on nanocrystalline titanium dioxide catalyzed photocatalytic degradation of aromatic carboxylic acids in aqueous medium. Journal of Industrial and Engineering Chemistry. 2011;**17**:331-339. DOI: 10.1016/j.jiec.2011.02.035

[25] Beltrán FJ, Rivas FI, Montero-de-Espinosa R. Catalytic ozonation of oxalic acid in an aqueous TiO_2 slurry reactor. Applied Catalysis B: Environmental. 2002;**39**:221-231. DOI: 10.1016/S0926-3373(02)00102-9

[26] Faria PCC, Órfão JJM, Pereira MFR. A novel ceria-activated carbon composite for the catalytic ozonation of carboxylic acids. Catalysis Communication. 2008;**9**:2121-2126. DOI: 10.1016/j.catcom.2008.04.009

[27] Peng J, Lai L, Jiang X, Jiang W, Lai B. Catalytic ozonation of succinic acid in aqueous solution using the catalyst of Ni/Al_2O_3 prepared by electroless plating-calcination

method. Separation and Purification Technology. 2018;**195**:138-148. DOI: 10.1016/j.seppur.2017.12.005

[28] Zhang T, Wiewei L, Croué J-P. A non-acid-assisted and non-hydroxyl-radical-related catalytic ozonation with ceria supported copper oxide in efficient oxalate degradation in water. Applied Catalysis B: Environmental. 2012;**121/122**:88-94. DOI: 10.1016/j.apcatb.2012.03.021

[29] Wang B, Xiong X, Ren H, Huang Z. Preparation of MgO nanocrystals and catalytic mechanism on phenol ozonation. RSC Advances. 2017;**7**:43464-43473. DOI: 10.1039/C7RA07553G

[30] Magallanes D, Rodríguez JL, Poznyak T, Valenzuela MA, Lartundo L, Chairez I. Efficient mineralization of benzoic and phthalic acids in water by catalytic ozonation using a nickel oxide catalyst. New Journal of Chemistry. 2015;**39**:7839-7848. DOI: 10.1039/C5NJ01385B

[31] Ikhlaq A, Brown D, Kasprzyk-Hordern B. Mechanisms of catalytic ozonation: An investigation into superoxide ion radical and hydrogen peroxide formation during catalytic ozonation on alumina and zeolites in water. Applied Catalysis B: Environmental. 2013;**129**:437-449. DOI: 10.1016/j.apcatb.2012.09.038

[32] Gan S, Lau E, Ng H. Remediation of soils contaminated with polycyclic aromatic hydrocarbons (PAHs). Journal of Hazardous Materials. 2009;**172**:532-549. DOI: 10.1016/j.jhazmat.2009.07.118

[33] Environment Agency United Kingdom. Risk Assessment of Naphthalene. 2007. Available from: https://echa.europa.eu/documents/10162/5f0beb6c-575f-4a1b-aff5-b37f06eb3852 [Accessed: 2018-01-10]

[34] Aguilar C, Rodríguez J, Chairez I, Tiznado H, Poznyak T. Naphthalene degradation by catalytic ozonation based on nickel oxide: Study of the ethanol as cosolvent. Environmental Science and Pollution Research. 2017;**24**:25550-25560. DOI: 10.1007/s11356-016-6134-2

[35] Oller I, Malato S, Sánchez-Pérez J. Combination of advanced oxidation processes and biological treatments for wastewater decontamination: A review. Science of The Total Environment. 2011;**409**:4141-4166. DOI: 10.1016/j.scitotenv.2010.08.061

[36] Moussavi G, Khavanin A, Alizadeh R. The integration of ozonation catalyzed with MgO nanocrystals and the biodegradation for the removal of phenol from saline wastewater. Applied Catalysis B: Environmental. 2010;**97**(1-2):160-167. DOI: 10.1016/j.apcatb.2010.03.036

[37] Poole J, Shi X, Hadad C, Platz M. Reaction of hydroxyl radical with aromatic hydrocarbons in nonaqueous solutions: A laser flash photolysis study in acetonitrile. The Journal of Physical Chemistry A. 2005;**109**(11):2547-2551. DOI: 10.1021/jp0452150

[38] Buxton G, Greenstock C, Helman W, Ross A. Critical review of rate constants for reactions of hydrated electrons, hydrogen atoms and hydroxyl radicals (•OH/O^-) in aqueous solution. Journal of Physical and Chemical Reference Data. October 2009;**17**:513-886. DOI: 10.1063/1.555805. 1988 Published Online

[39] Wu K, Zhang F, Wu H, Wei C. The mineralization of oxalic acid and bio-treated coking wastewater by catalytic ozonation using nickel oxide. Environmental Science and Pollution Research. 2018;**25**:2389-2400. DOI: 10.1007/s11356-017-0597-7

[40] Verlicchi P, Aukidy M, Zambello E. Occurrence of pharmaceutical compounds in urban wastewater: Removal, mass load and environmental risk after a secondary treatment: A review. Science of the Total Environment. 2012;**429**:123-155. DOI: 10.1016/j.scitotenv.2012.04.028

[41] Farre M, Ferrer I, Ginebreda A, Figueras M, Olivella L, Tirapu L, Vilanova M, Barcelo D. Determination of drugs in surface water and wastewater samples by liquid chromatography–mass spectrometry: Methods and preliminary results including toxicity studies with *Vibrio fischeri*. Journal of Chromatography A. 2001;**938**:187-197. DOI: 10.1016/S0021-9673(01)01154-2

[42] Kanakaraju D, Motti C, Glass B, Oelgemoller M. TiO_2 photocatalysis of naproxen: Effect of the water matrix, anions and diclofenac on degradation rates. Chemosphere. 2015;**193**:579-588. DOI: 10.1016/j.chemosphere.2015.07.070

[43] Jallouli N, Elghniji K, Hentati O, Ribeiro A, Silva A, Ksibi M. UV and solar photo-degradation of naproxen: TiO_2 catalyst effect, reaction kinetics, products identification and toxicity assessment. Journal of Hazardous Materials. 2016;**304**:329-336. DOI: 10.1016/j.jhazmat.2015.10.045

[44] Ghauch A, Muthanna A, Kibbi N. Naproxen abatement by thermally activated persulfate in aqueous systems. Chemical Engineering Journal. 2015;**279**:861-873. DOI: 10.1016/j.cej.2015.05.067

[45] Chen C, Yan X, Yoza BA, Zhou T, Li Y, Zhan Y, Wang Q, Li QX. Efficiencies and mechanisms of ZSM5 zeolites loaded with cerium, iron, or manganese oxides for catalytic ozonation of nitrobenzene in water. Science of the Total Environment. 2018;**612**:1424-1432. DOI: 10.1016/j.scitotenv.2017.09.019

[46] Tang Y, Pan Z, Li L. pH-insusceptible cobalt-manganese immobilizing mesoporous siliceous MCM-41 catalyst for ozonation of dimethyl phthalate. Journal of Colloid and Interface Science. 2017;**508**:196-202. DOI: 10.1016/j.jcis.2017.08.017

[47] Li C, Jiang F, Sun D, Qiu B. Catalytic ozonation for advanced treatment of incineration leachate using (MnO_2-Co_3O_4)/AC as a catalyst. Chemical Engineering Journal. 2017;**325**:624-631. DOI: 10.1016/j.cej.2017.05.124

Chapter 3

Ozone Dosage is the Key Factor of Its Effect in Biological Systems

Tatyana Poznyak, Pamela Guerra Blanco,
Arizbeth Pérez Martínez, Isaac Chairez Oria and
Clara-L. Santos Cuevas

Additional information is available at the end of the chapter

http://dx.doi.org/10.5772/intechopen.76843

Abstract

The applications of ozone are not only restricted to environmental remediation or industrial areas. This gas has been applied in medicine to treat several diseases, where positive effects have been confirmed by many clinical studies. According to the European Medical Society of Ozone and the National Center of Scientific Investigation in Cuba, it has not been possible to validate ozone's effectiveness by traditional analytical methods. Thus, this investigation proposed evaluating the effect that ozone has on biological substrates (murine models with induced carcinogenic tumors, inflammation, and wounds), studying the variations that ozone (dissolved in physiological solution or ozonated vegetable oils) provokes over the total unsaturation of lipids (TUL), and by using the so-called method double bond index (DB-index), make a correlation with the dynamic reactions obtained by several analytical methods according to each experimental stage considered in this study.

Keywords: ozone therapy, cancer, ozonated oils, inflammation, wound healing, total unsaturation, double bond index

1. Introduction

Ozone is a gaseous molecule formed by three oxygen atoms; it has a blue color (when dissolved in water) with a strong acrid aroma and a molecular weight of 48 mg/mole. The ozone molecule has a cyclic structure with a distance between atoms of 1.25°A. It has a solubility of 49/100 ml of water (at 0°C), that is 10 times greater than the oxygen solubility (4.89/100 ml of water) [1].

Ozone concentration on air (mg/L)	Symptomatology
0.1	Respiratory tract and eyes irritation
1.0–2.0	Rhinitis, cough, headaches, nausea and vomiting
2.0–5.0 (10–20 min)	Dyspnea and bronchial spasms
5.0 (60 min)	Acute pulmonary edema and occasionally respiratory paralysis
10.0	Death in 4 h
50.0	Death in minutes

*The ozone toxicity on the respiratory system should not be extrapolated to the circulatory system due to the differences in biochemistry and the metabolic regimen [3].

Table 1. Toxic effects of ozone in gaseous phase*.

The excessive emissions of NO, NO2, CO, CH4, H2SO4, among others, have favored the increase in ozone concentration in the tropospheric space, above 0.1 ppmv. The reactions between the chemical compounds abovementioned give rise to the so-called photocatalytic smog, which has become the main toxic substance for lungs, eyes, nose, and skin. There are several symptoms that could appear according to ozone concentration that exists in the air (**Table 1**) [1, 2].

2. Ozone in medicine

According to the European Ozone Medical Society and the National Center for Scientific Research in Cuba, among others, the following diseases can be treated with ozone: abscesses, acne, AIDS, allergies, anal fissures, arthritis, asthma, cancerous tumors, cerebral sclerosis, problems in the circulatory system, cirrhosis of the liver, corneal ulcers, cystitis, diarrhea, fistulas, boils, gangrene, gastric ulcers, intestinal disorders, glaucoma, hepatitis, herpes, hypercholesterolemia, colitis, mycosis, and osteomyelitis [2].

Unlike research on the application of ozone at the industrial level, studies in the medical field are scarce. Moreover, the studies describing the interactions of ozone with substances of biological origin and their kinetic implications have not described the entire reaction mechanisms. One of the most important premises in the application of ozone for medical aspects establishes the induction of an extraordinary and temporary response of the body systems associated with peroxidation of lipids and the antioxidant scheme of the organism [1–3].

However, it is suspected that the oxidative effect of ozone causes different effects on the immune system, sympathetic and parasympathetic systems among others. It is well known that the presence of compounds derived from oxidation reactions in the human body produces a cascade of biochemical reactions that has been clearly explained but not associated with the presence of ozonation-derived byproducts. This condition occurs in many events that compromise the health of the human being, such as deep wounds, appearance of neoplasms, and so on. However, the mechanism of reactions through which the cascade of biochemical reactions occurs is not entirely known with certainty [3–5]. Even though it is accepted that ozone (under the adequate dosing strategy) produces a significant number of benefits in the human body because it is dissolved in oxygen, it increases oxygenation in the blood, improves circulation, stimulates oxygenation in tissues, and so on, it has not been established what are the mechanisms that generate such important advantages from the clinical point of view [1, 2].

3. How ozone acts and how its toxicity can be avoided

Oxygen is essential for life; nevertheless, this gas has long-term negative effects. Reactive oxygen species (ROS) are formed during cellular respiration. The hydroxyl radical $OH^{\cdot}$ is the most destructive ROS for enzymes and deoxyribonucleic acid (DNA). The aging process and metabolic disorders (arteriosclerosis, diabetes, cellular degeneration, etc.) can be worsened by the presence of ROS. The application of an excessive ozone dose used in medical therapy may aggravate the ROS effect on the body. This process can be prevented if there are proper control methods in ozone dosage, regardless of the medical ozone technique [1].

Notice that, in the ozone-oxygen mixture, the former is not equilibrated with ozone, because ozone reacts immediately with a certain number of molecules in biological fluids, mainly antioxidants, proteins, carbohydrates, and specifically the polyunsaturated fatty acids (PUFAs) [3].

The reaction kinetics and the sequence of such reactions are uncertain, and it has been briefly described [1, 6–8]. The subsequent formation of byproducts that may be responsible for the clinical effects of ozone and the accumulation of final products must be controlled in order to avoid some of the undesirable side effects of the therapies based on ozone.

It is widely accepted that the main reactions of ozone with biological molecules are executed according to the following stages [1]: (1) ozone reacts with ascorbic acid, uric acid, sulfhydryl groups (SH−) from proteins, and glycoproteins generating ROS, which trigger several biochemical stages in the blood ex vivo. The ROS are neutralized 0.5–1.0 min later by the antioxidants of the system and (2) ozone reacts with the double bonds (>C=C<) of arachidonic acid and triglycerides in the plasma, which produces a molecule of hydrogen peroxide (H_2O_2) and two aldehyde molecules known as lipid peroxidation products (pPOL).

According to these stages, it is possible to claim that not ozone but ROS and pPOL are the compounds responsible for the multiple biochemical reactions that occur in the cells of the body, in particular, in particular, the second reaction, which has been characterized as the key factor of the therapeutic effects of ozone. In this way, the study of the ozonation byproducts formation improves the understanding of the clinical effects, which is helpful to choose the better ozone's application scheme in a medical treatment.

As soon as ozone dissolves in biological fluids, it reacts with PUFAs, and then the concentration of hydrogen peroxide increases. However, with a similar rate, it begins to diminish as the molecule diffuses quickly toward erythrocytes, leukocytes and platelets, while several antioxidative processes are performed. Due to the presence of enzymes such as GSH-Px and GSH, the intracellular concentrations of hydrogen peroxide are reduced within the plasma and the intracellular fluid [1, 3].

The activity of the pPOL, under prolonged therapies, can give rise to the regulation of antioxidant enzymes, the appearance of oxidative proteins, and the release of stem cells, which is considered a crucial factor to explain some effects of ozone applications as medical therapy [1].

4. Ozone application pathways

The therapeutic indications of ozone are based on the theory that at low concentrations of this gas (in the gaseous phase), some significant phenomena occur within the cell. It has been proved that at concentrations of 5–10 mg/L or lower, there are therapeutic effects with a wide margin of safety in the patient. At present, concentrations ranging from 5 to 60 mg L^{-1} are accepted for the medical application of ozone [9].

4.1. Direct methods

Rectal insufflation: The gaseous mixture is introduced to human body by the rectum, and it is absorbed in the bowels.

Intramuscular injection: In this technique, 10 mL of gaseous mixture are injected in the buttocks of the patient.

Major and minor autohemotherapy: This technique has been used since 1960. The minor autohemotherapy requires 10 mL of blood, which is put in contact with the gaseous mixture to finally return it to the human body. The major autohemotherapy requires 50–100 mL of blood, which is put in contact with the gas for a few minutes to then return it to the patient.

Ozone bag: A plastic bag is placed around the treated area. The gaseous mixture is pumped into the bag, then the gas is absorbed by the human body through the skin. This is one of the methods where the reaction occurs in two stages, one is absorption by the skin and the second is the direct reaction of the skin compounds with ozone. That implies a combined model that involves mass transfer of ozone and its reaction with the bio-substrate [1, 2, 9].

4.2. Indirect methods

Ozonated water: Ozonated water is used to wash wounds, skin burns and skin infections.

Intra articular injection: The ozonated water is injected directly between the joints and is used for the treatment of arthritis and rheumatism.

Ozonated oil: The ozonated oil is applied as an ointment for long times with low ozone doses [1, 2, 9].

http://dx.doi.org/10.5772/intechopen.76843

5. Control methods for ozone's therapeutic applications

Oxidative stress is the main concept that explains ozone's therapeutic effect over the human body. Ozone's paradoxical effect as a promoter of antioxidant response capable of regulating oxidative stress is common in the animal kingdom. This fact suggests that the adequate ozone's dose, besides the oxidation induced in biological subjects, may enhance the antioxidant response of the living organism. This is a critical factor issued by the immunological system to overcome infections, ischemia, and cellular regeneration [3].

Most of the technical methods employed to control the medical application of ozone are based on the measurement of oxidative stress. Some of the typical methods employed to measure the oxidative stress include the quantification of reactive species by electronic paramagnetic resonance; the analytical determination of antioxidants and measurement of total antioxidant capacity; the detection of oxidized biological markers, such as the lipid peroxidation products (pLPO), malondialdehyde, 4-hydroxynonenal, isoprostane, and oxidized proteins; as well as the measurement of damaged DNA [10]. While these methods measure key species associated with the oxidative stress, they present some inconveniences regarding their own analytical limitations, mainly the sensitivity and the time required to complete the analysis.

Nowadays, ozone's medical application remains empirical. According to the "Madrid Declaration on Ozone Therapy," each patient responds differently to the controlled oxidative stress induced by ozone treatments; thus, ozone's administration must be developed in a progressive way, that is, starting with small doses and progressively increasing them [9].

5.1. Determination of the total unsaturation

The determination of the total unsaturation (TU) of organic compounds is a technique developed by Russian researches in the middle of the last century [11]. It is a useful tool based on ozone's reactions, particularly, with the double bonds (>C=C<). Ozone reacts selectively with different compounds; one of the most specific reactions of ozone takes place with unsaturated organic compounds [12]. The reaction rate constants of ozone with all the >C=C < bonds are similar, regardless of the structure of the compounds that contain them [13]. Through the TU technique, it is possible to determine the TU of the lipids in the biological substrates, and the ones contained in vegetable oils, due to their composition, which make TU a suitable technique to control ozone's therapeutic applications.

By this technique, the ozone reactive substrate contained in one sample can be quantified in a precise way (±1%) and in a short time of analysis (1–3 min) [11, 14, 15]. The TU determination consists of the measurements of ozone necessary to react with certain samples diluted in chloroform. Afterwards, this quantity is compared with ozone consumed in its reaction with a standard sample of known concentration, as well as the stoichiometric of its reaction with ozone. Ozone's consumptions are obtained from the area of the characteristic plot of ozone concentration versus time, called ozonogram. The detailed procedure can be reviewed in [15, 16]. The mathematical formula used to calculate TU is:

$$TU = \frac{C_{ST} \times V_{ST} \times S_S \times V_{SOL}}{S_{ST} \times V_S \times W_S} [=] \left[\frac{mol_{>C=C<}}{g_{sample}} \right] \quad (1)$$

where C_{ST} is the concentration (mol L^{-1}) of the standard solution, V_{ST} and V_S are the volumes (ml) of the standard and the sample, respectively, S_{ST} and S_S are the ozonogram areas for the standard and samples, respectively, while V_{SOL} and W_S are the solution volume (ml) and weight of the sample (g), respectively.

5.1.1. Determination of the ozonation degree of vegetable oils by TU

Vegetable oils are composed mainly of fatty acids (free and as esters in triglycerides); the unsaturated ones constitute the principal substrate that reacts with ozone. In addition, vegetable oils contain minor compounds (unsaponifiable matter), which include sterols, polyphenols, pigments, antioxidants, as well as characteristic compounds extracted from seeds. These compounds are also reactive with ozone because they contain double bonds and some other oxidable elements. The products of ozonated oils constitute a mixture of peroxides (isoozonides, hydroperoxides, poly peroxides) with therapeutic action. The formation of these species is described by Criegee mechanism. The type and their yield depend on reaction conditions [12, 13, 17, 18]. **Figure 1** shows the reaction pathway of ozone with the unsaturated compounds of oils.

TU quantifies ozone mass that reacts with an oxidable sample. Notice that this method considers that the stoichiometry of the reaction of ozone with >C=C < is 1:1. Then, the TU quantifies the oxidizable substrate by ozone that is contained in the analyzed sample. Thus, the ozonation degree of oils can be easily obtained, as the percentage of all compounds in oils that react with ozone (major and minor compounds). The importance of a reliable determination of the ozonation degree is related to the therapeutic effects of ozonated oils, which, in turn, strongly depend on the type of oil and its ozonation degree.

5.1.2. Double bond index (DB index)

The DB index is the term used to extend the TU application to biological substrates. It is obtained from the measurement of ozone that has reacted with the double bonds of lipids, previously extracted from biological fluids or tissues [15]. With the determination of the total unsaturation of lipids in plasma and cell membranes, it is possible to evaluate changes in

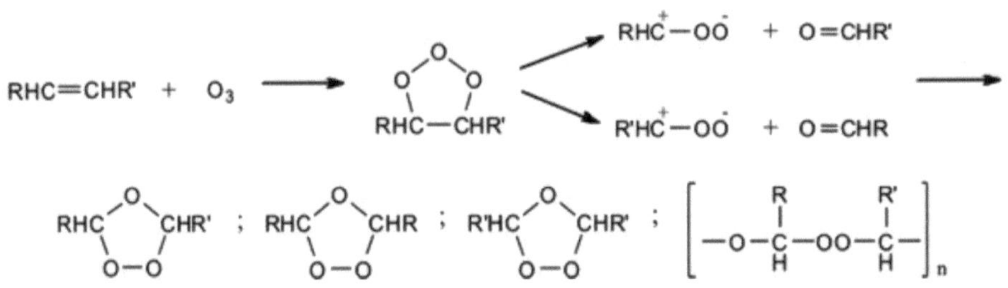

Figure 1. Reaction pathway of ozone with unsaturated compounds.

lipid metabolism [11, 14]. The DB index is strongly related with the level of oxidative stress in subjects, since the lipids involved in the measurement correspond to those remaining after the oxidative stress mechanism [15].

The DB index determination corresponds to the procedure described for TU measurement. For liquid samples, such as blood plasma, the mathematical expression DB index is calculated according to:

$$\mathrm{DB-index} = \frac{C_{ST} \times V_{ST} \times S_S \times V_{SOL}}{S_{ST} \times V_S \times V_{pi} \times K}[=][c.u] \tag{2}$$

where C_{ST} is the concentration of the standard solution (mol L^{-1}), V_{ST}, and V_S are the volumes (ml) of the standard and the sample, respectively, S_{ST} and S_S are the ozonogram areas for the standard and samples, while V_{SOL} and V_{PL} are the volume (ml) of solution and the volume (ml) of blood plasma from sample, respectively; K is a correlated coefficient equal to 10^{-7} mL/mole. C.u means "conditional unit" (1 c.u = 1×10^{-5} mole D.B./mL) [15].

The DB index, measured in healthy subjects, has been determined showing its impendence of age or sex, excluding children ≤1 year and aged people ≥60 years. Apparently, only diseases or pathologies produced changes in this value [14]. The DB index of lipids (blood plasma and erythrocytes) for healthy European and Mexican population are listed in **Table 2**.

Some reported clinical cases illustrated the prognostic and diagnostic criteria of the DB index changes. Among the diseases where the DB index has been successfully used as a preclinical tool, the oxidative process of pneumonia in children was precisely characterized by this method [11]. In this study, and inverse correlation between the DB index and the LPO activity was observed. The authors concluded that the DB index evolution can describe the disease evolution accurately [11].

Another relevant case of the DB index application in medical procedures corresponds to its application to evaluate the therapy effectiveness of burned patients [19]. Depending on the burned magnitude, the reported DB index values of different patients ranged from 34 to 287c.u. The evolution of the DB index correlated for each subject with its personal damage and its treatment effectiveness. The authors concluded that the changes in the DB index were observed before the clinical manifestations, showing its potential as a prognostic tool for clinical practice [19]. Some other diseases where the DB index has been related with the evolution of the illness and the effect of the treatment include cancer [20], diabetes [14], exposition to hexavalent chromium [21], as well as inflammatory processes [22]. In these cases, the DB index resulted in a powerful tool to adjust the therapeutic treatment, according to the individual needs of the subjects under treatment.

Population	**DB_{plasma}–index (c.u)**	**$DB_{erythrocytes}$–index (c.u)**
European	250 ± 10	50 ± 2
Mexican	160 ± 10	100 ± 2

Table 2. DB-index value for European and Mexican healthy population [15].

6. In vivo studies of the TU and the DB index application

6.1. Direct applications: cancer

The methodology proposed in this section evaluated the implantation of C6 cells in an animal (murine) model [23, 24]. The oxygen and the ozone dissolved in the saline solution were dosed by an intraperitoneal pathway in athymic mice (Balb/CNu/Nu). To evaluate the effect of ozone dosage on the tumor implanted in mice, the measurements of the DB index were carried out on the lipid fraction of plasma of blood, erythrocytes, tumor, and liver.

Ozonated physiological solution (NaCl 0.9%): Physiological solution (0.9% NaCl) was used as carrier media for the oxidant agent (ozone or oxygen). The ozone concentration in oxygen was 4.6 ± 0.2 mg L^{-1} that corresponds to its concentration of 1.15 ± 0.2 mg $^{L-1}$ of ozone in the saline solution. Considering the volume of the injected physiological solution (90 μL), only **0.103 μg** of ozone was injected to mice in treatments.

Therapeutic protocol: This experiment considered four groups (n = 6) of athymic nude mice with C6 glioma that practically have the same tumor size (74.60 ± 21 mm^3). The oxidant agents were dissolved in the physiological solution, and they were administrated into the mice by intraperitoneal injection (90 μL) [25]. The treatment period length was 15 days. The number and the frequency of injections were different for each treated group. In the first and second groups, the injections (oxygen for the first and ozone for the second) were carried out every second day (7 times); the third group was treated with ozone every fifth day (three times). Then the mice were sacrificed, and the samples of blood and tissues were selected to determine the DB index.

Evolution of volume and necrosis of tumor: The variation of tumor volume for 15 days is shown in **Figure 2**. As we can see, ozone promoted the tumor volume growth compared with the control group by 10 and 44% every fifth and second day, respectively, both compared with the control group. On the contrary, oxygen inhibits the tumor growth by 30%. Even when the tumor volume results suggest a better performance in the group treated with oxygen, the tissue necrosis demonstrate a lower activity of tumor cells in groups treated with ozone. Furthermore, the microscopically obtained results showed that the ozone dose influenced the tumor necrosis.

Some studies have shown that oxygen may inhibit the tumor angiogenesis, which limits the nutrient availability [26], and could be related with slower tumor growth observed in the group treated with oxygen. On the other hand, ozone showed a stressing effect, which was reflected by the accelerated tumor growth. Ozone may induce a pronounced influence on tumor metabolism, particularly in the respiratory cycle and glycolysis, showing a positive influence on oxygen utilization in tumor [27]. These facts may explain the increased rate of tumor growth [28, 29].

Since tumor growth was slower when the ozone dose also was applied every fifth day (compared with ozone applied every second day), and a higher necrosis was observed, we may conclude that ozone dose plays a major role in two observed phenomena, in regulating the

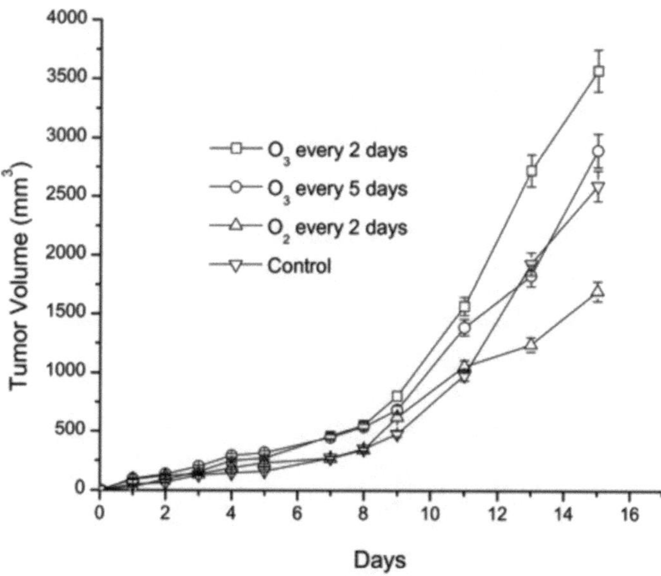

Figure 2. Tumor volume increase associated with the dosage strategy showing the control group, pure oxygen and ozone dosed every second and every fifth day (n = 6).

rate of the tumor growth and in the tissue necrosis. It is important to note that both, the tumor cell activity and tumor necrosis, the significant positive effects of the treatment were achieved under the smaller ozone dose. Considering that tumor necrosis is a positive result of the treatment based on ozonated saline solution, the smaller ozone dose (every fifth day) was the better tested strategy to treat this type of tumor.

[18F] FDG in tumors with PET/CT: Tumor metabolic activity: [18F] FDG (2-deoxy-2-[18F]-fluoro-D-glucose)-positron emission tomography (PET) and X-ray CT imaging were performed using a micro-PET/CT scanner (Albira, ONCOVISION, Spain). **Figure 3** shows the variation of FDG in the tissue of tumor that was obtained by the image processing analysis corresponding to the set of three planes of exposition (top of the image). In the center of the figure, the acquired PET images (with the gamma camera taken in the same planes) are located. At the bottom, the over-position of both images, tomographic and PET, is shown to correlate the tumor anatomical position with its activity.

Figure 4 represents the specific areas that demonstrate the tumor activity by color variation. The red color corresponds to larger tumor activity, and, contrary to that, the blue areas describe the regions with smaller or even null activities [30–33]. **Figure 4b** corresponds to the mice dosed with dissolved oxygen, showing an area in red color that is larger than the one detected for the mice of the control group. **Figures 4c** and **d** represent images of the mice dosed with ozone every second and fifth day, respectively. As we can see, under the smaller ozone dose the significant decrease of tumor cell activity is obtained (>80%). It is important to note that in both cases, the tumor cell activity and tumor necrosis, the significant positive effects of the treatment were achieved under the smaller ozone dose.

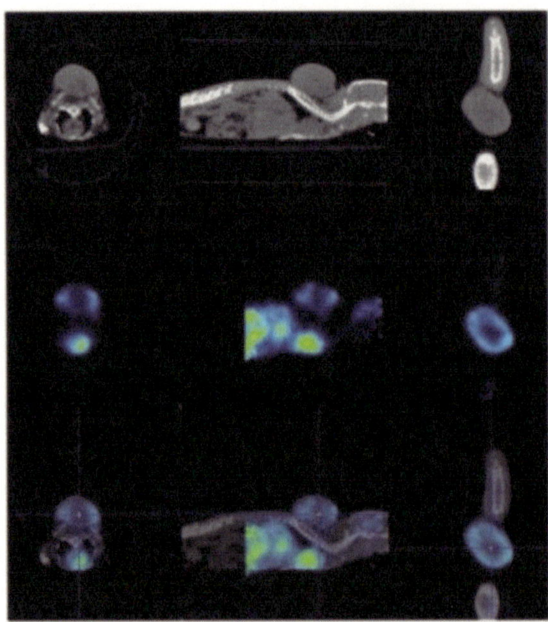

Figure 3. Example of microPET image obtained when the ozone gas Dosage is administered every fifth day.

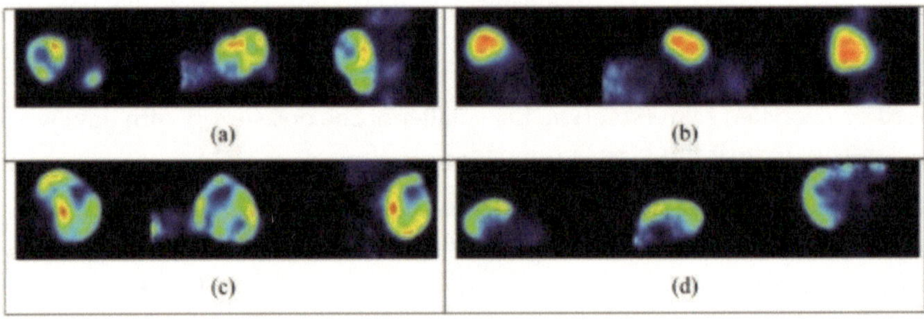

Figure 4. 18FDG tumor activity of the considered studied group control (a), only oxygen every second day (b), ozone every second day (c) and ozone every fifth day (d).

The DB index variation of plasma, erythrocytes, tumor, and liver: The measurement of DB index (reactive sites of biological substrates) of lipids of plasma and erythrocytes as well as of tumor tissue and liver was carried out to find the possible correlation with the ozone dose and its effect on the tumor volume and activity. **Figure 5** depicts the DB index of the plasma and erythrocytes from mice after the ozone and oxygen treatment, both compared with the control group. The first fact to note is the values of the DB index of both plasma and erythrocytes, which are very high in comparison with healthy mice: 2.7×10^{-2} and 2.35×10^{-2} with respect to 2.0×10^{-5} and 0.57×10^{-5} mol ml^{-1}. Usually the values of the DB index of lipids of the plasma and erythrocytes are close to each other (0.57 and 0.43×10^{-5} mol ml^{-1}) [15]. These high values of the DB index in mice with this type of cancer indicate that lipid peroxidation was substantially reduced.

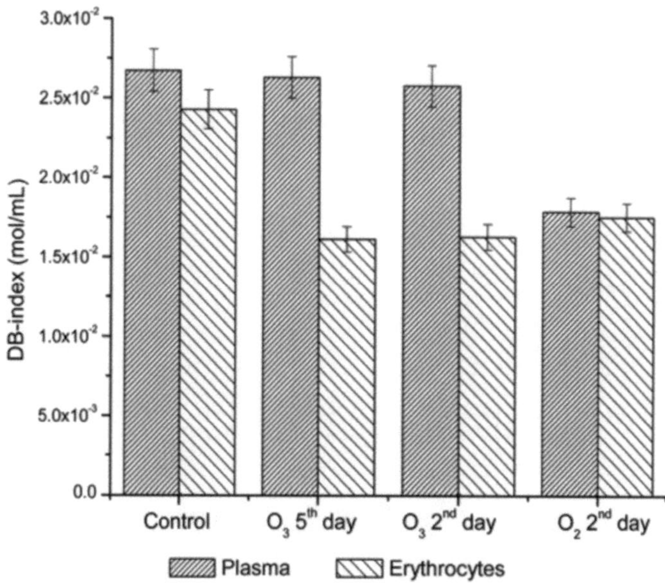

Figure 5. DB-index values obtained from lipid samples extracted from plasma and erythrocytes (n = 6).

On the other hand, in mice treated with ozone and oxygen, this index for erythrocytes was lower than the control group (around 32%), which indicates the tendency to normalization of LPO.

Figure 6 depicts the DB index of lipids extracted from tumor and liver tissues. These were measured in our previous study for the first time in mice with tumors [20]. The DB index of tumors reduces by 83% and of the liver by 70% in the group of mice with smaller doses of ozone compared to the control group and the other groups treated with ozone and oxygen. This phenomenon was observed in both tumor and liver tissues and it seems to be a consequence of the modification in metabolism promoted by ozone. According to the preliminary studies, the cancer cells repressed the TIGAR enzymes. The lower concentrations of these enzymes keep the ROS concentration at an abnormal higher level [8] that caused the apoptosis of cancer cells [34]. The mice treated with ozone every second day and oxygen had higher values for DB index of tumor tissues, which is related with the cell activity observed microscopically. In fact, there was no significant statistic difference of the DB index measured in tumor tissue treated with ozone every second day and the control group. It seems that this dose had no effects over the tumor cell metabolism, contrary to the lower dose (every fifth day), where the mice's system was able to better regulate the oxidative stress induced in the treatment, and this was reflected in both the lower cell activity and the tissue DB index. In the group of mice treated with oxygen, the DB index increased 33% compared with the control group. This may be caused by the overexpression of some enzymes, such as FAS, which improved the cell proliferation by oxygen presence [9, 20].

However, the DB index of liver tissue increased by 50% compared to the control group. This variation points out to the sensibility of this index on the ozone dose, that is, this value decreased by 70% in mice dosed every fifth day and increased by 50% in subjects dosed every

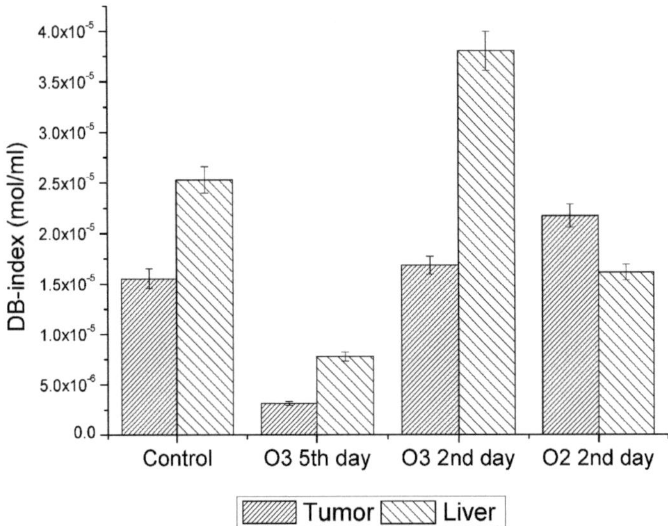

Figure 6. DB-index values observed in tumor and liver tissues (n = 6).

second day. In the group treated with oxygen, the DB-index in liver decreased 40% compared with the control, due to the decrease of the consumption of the energy caused by the decrease of tumor volume. The DB index of the liver suggests a modification in the energy consumption associated with the tumor activity because of the treatment. The accumulation and production of lipids appeared to be inhibited in mice treated with ozone (every fifth day) and oxygen [25]. This effect could be explained by two possible metabolic pathways of the fatty acids synthesis in the organism. The first considers that lipids in plasma come from energetic reserves of the liver (glucagon), which is regulated by the lipid reduction in blood indeed. The second assumes that the fatty-acid cycle regulates the lipid concentration in plasma, which is also regulated by the liver, but there is no energetic transformation. Under the higher dose of ozone, the liver regulates the lipid imbalance to compensate the oxidized fatty acids. This additional lipid source may explain the increase of the DB index.

On the other hand, under the smaller ozone dose, the lipid-accumulation effect was not observed. The last can be a result of the regulatory process conducted outside of the liver tissue that seems to justify the DB index reduction. The tumor cell activity correlated with the DB index of the lipids is obtained from the tumor tissue. When their activity was the smallest or almost zero, the DB index value was smaller also. This confirms that the DB index determination can be a reliable method to control the medical treatment efficiency for regulating the tumor growth and its activity as well.

6.2. Indirect applications: vegetable oils

The ozonated vegetable oils (OVO) have shown interesting applications in diverse fields, such as food, pharmaceutical, and cosmetic industries, since their applications have resulted in several positive in vitro, in vivo, and clinical effects. In addition to their therapeutic potential, the OVOs have some advantages over other ozone applications, since they are composed of stable reaction products [35]; thus, it is not necessary to produce them in situ. This is an additional advantage from a commercial point of view.

Among the most reported therapeutic effects of the OVO, one may list bactericidal, fungicidal, as well as inflammation and wound-healing mediators [35–38]. These effects are highly related with the oil type, as well as the ozonation degree.

The determination of the TU and DB index has been useful in the application of vegetable oils. By using these parameters, it is possible to control the ozonation conditions to achieve a certain ozonation degree, as well as observe the treatment's evolution and evidence of the biochemical changes derived from the treatment.

6.2.1. Ozonation degree of vegetable oils

The therapeutic action of the OVO depends on the accumulated ozonation products. A lot of techniques have been employed to characterize these compounds, such as the spectroscopic methods (Fourier transform infrared (FT-IR) and hydrogen-1 nuclear magnetic resonance (^{1}H NMR). The identified products by these techniques corresponding to those described by the Criegee's mechanism (iso-ozonides, poly-peroxydes, hydroxyperoxides) [12, 13, 17, 18, 39, 40].

Different studies have justified that the observed effects of the applied ozonated oils depend on the oil's ozonation degree. For example, the in vitro tests showed that the bactericidal and fungicidal effects increase when the ozonation degree increases [41–43]. Complementary, the in vivo evaluations showed that the adequate ozonation level depends on the treated illness and the vegetable source of oil [44–46].

For example, the inflammation process induced in the mice's skin by 2,4-dinitrofluorobenzene was inhibited after the application of olive oil ozonated 100% (iodine value = 0). However, the repeated applications produced hair losses, hypervisibility and swelling reactions [44]. Another work demonstrated that the ozonated sesame oil showed diverse effectiveness for mice's wound healing, depending on the peroxide index of the applied oil [45]. The authors found that the better peroxide value was 1631±64 mEq/kg. The higher and lower values of oil's peroxide value were less effective.

In our previous work, we studied the ozonation of two oils: sunflower (SFO) and grape seed (GSO) [16]. The different ozonation products (mainly ozonides) were identified by the spectroscopic techniques (FT-IR, ^{1}H NMR). Also, the changes in OVO's viscosity were associated with the formation of poly-peroxides. We also determined the dynamics of >C=C < decomposition and product accumulation. We found that the TU decrease was similar in both oils, but the distribution of their ozonation products was different. It was established that the maximum amount of ozonides were formed faster in GSO. This oil accumulated a higher proportion of poly-peroxides related with its viscosity, when compared with the SFO [16].

Since the therapeutic effects of ozonated oils are strongly related to the accumulation of the ozonation products, our previous investigations offer an alternative method for controlling the ozonation degree in the preparation of ozonated oils.

6.2.2. *Inflammation and wound healing*

The effectivity of TU and DB index in ozonated oils' applications was evidenced in our previous work [22], where the anti-inflammatory and wound healing effect of the ozonated grape seed (GS) and sunflower (SF) oils in mice were tested (for wound healing, diabetic mice were tested). The ozonation degree of both oils (determined by TU) was related with the in vivo effect of oils. For comparison, the ozonated physiological solution was applied (subcutaneous injection, only for inflammation test), as well as the commercial drugs indomethacin and Furacin® for anti-inflammatory and wound-healing tests were used, respectively.

Some differences on the biochemical effect of different treatments were found, depending on the oil type and their ozonation degree. These differences were revealed by the DB index values of the treated tissues [22].

In the case of the SF oil with the ozonation degree of 44%, the DB index of 2.32×10^{-4} mol g^{-1} corresponds to the inflammation inhibition (INI) which is about 32%. For the GS oil, the maximum INI was 25% under the ozonation degree of 24% and corresponds to the DB index of 2.26×10^{-4} mol g^{-1}. In this last case, the increase of the ozonation degree of GS oil up to 41% decreases the INI down to 23%.

The effect of the vegetable source of oils on both the INI and the DB index suggests an active participation of their minor compounds typically contained in oils, in the response of the immunological system (polyphenols, tocopherols, carotenoids, chlorophylls). **Figure 7a–d** shows the chemical structures of some these compounds. As seen, they are susceptible to reacting with ozone, due to their unsaturated structures. So, they may participate in the therapeutic effects of ozonated oils. Even when the concentration of these compounds is low, in comparison with unsaturated fatty acids, they can consume considerable amounts of ozone. For example, the stoichiometry of the reaction of phenols with ozone is 1:3.5, while >C=C< with ozone is 1:1.

A lower INI and the DB index of the pavilion of the ear (15% and 1.57×10^{-4} mol g^{-1}, respectively) resulted from the application of ozonated physiological solution (PS), when compared with ozonated oils. This fact indicates that the action mechanism of ozone, when applied directly (dissolved in PS) or indirectly (ozonated oils), is different.

In the case of wound-healing evaluation for diabetic mice, we also found promising results, considering that this sickness negatively affects the wound-healing process. We observed that the complete wound healing of diabetic mice treated with ozonated oils was obtained during 12 days [22]. This time is comparable with the one reported for wound healing of non-diabetic mice (14 days) treated with ozonated sesame oil [45]. It is worth mentioning that no infection signs were observed over the mice skin tissue. The glucose content was also monitored throughout the treatment. None of the agents showed a regulatory effect of glucose, as expected. Then, the cutaneous application of ozonated oil did not have a systemic effect which is considered a positive effect because these oils acted locally.

Our results showed that minor compounds presented in oils may have biological activity, which contributes to the effect of the well-known therapeutic ozonation byproducts, namely

(a)

(b)

(c)

(d)

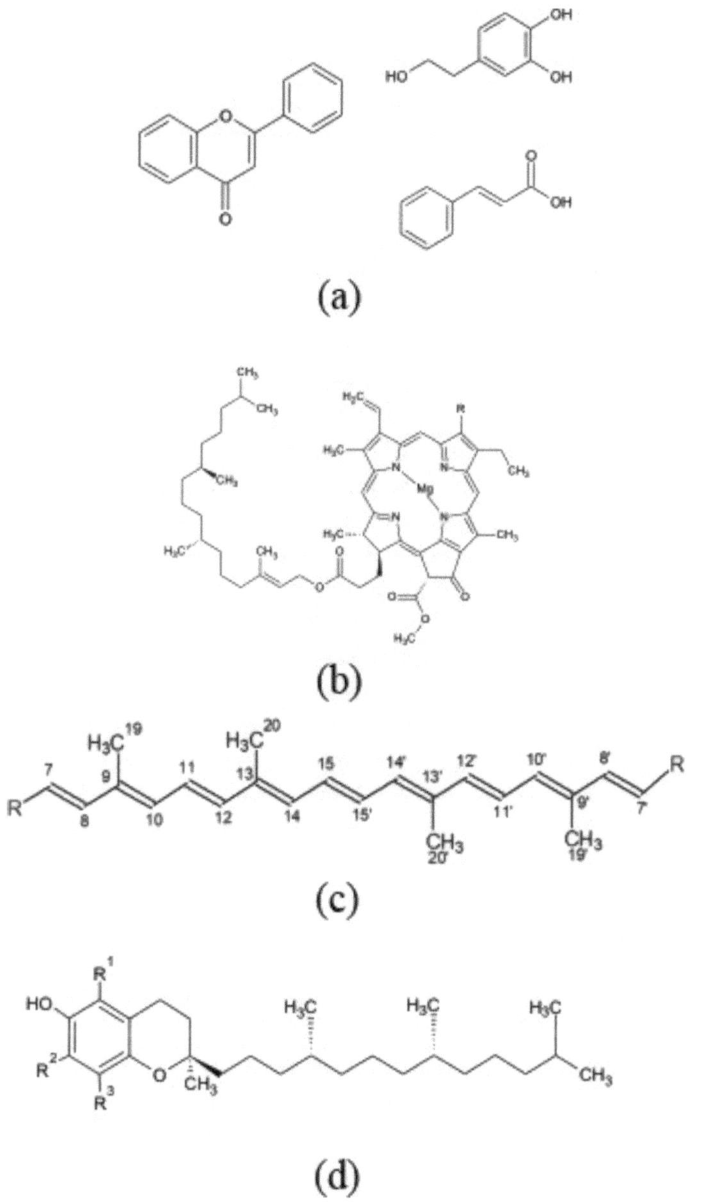

Figure 7. (a) Structures of polyphenols. (b) Structures of chlorophylls. (c) Structures of carotenoids. (d) Structures of tocopherols.

ozonides, as a product of triglycerides' ozonation. This effect was more pronounced in the case of inflammation. For wound-healing tests, a slight improvement was observed in the implementation of the SF oil, compared to the GS oil (both ozonated up to the same ozonation degree). Based on these data, we may conclude that the positive clinical effect of ozonated oils depends on their ozonation degree and their nature and then the composition of their minor compounds.

7. Conclusion(s)

The TUL determination was an adequate parameter to evaluate the effect of ozone on biological substrates. The versatility of this technique allowed the control of the ozonation degree of oils, as well as the correlation of biochemical changes in tissues involved in ozone-based treatments for C6 tumor cells, inflammation, and wound healing, considering direct and indirect applications. The effect of dissolved ozone dosage (direct application) on the tumor evolution was observed, and the main result was that at low doses of ozone (every 5 days), there is a greater effect on the inactivity of C6 glioma cells, decreasing their reproduction and therefore, reducing the DB index of tumor tissues, in comparison with other groups. This effect depends on the type and stage of the disease. Since it has been reported that the application of ozone reduces the size of certain tumors, in this context, it was observed that although ozone had a positive effect with respect to the activity of the tumor quantified by micro PET and the DB index determination, the volumetric growth of the tumor was disproportionated. The results presented in this study demonstrated that the key factor for controlling the tumor activity, inflammation, and wound healing through direct and indirect applications of ozone was precisely ozone's dose. Our results suggest that low doses of ozone may induce a micro-oxidative stress that stimulates the organism to perform a redox regulation, which is reflected as a self-inhibition of the cancer tissue activity. The micro-dose of ozone may have a systemic and prolonged effect on the organism. Therefore, in this study, three injections for 15 days were enough to get a decrease, > 80%, of tumor cell activity in mice. In addition, the DB index values pointed at different reaction mechanisms of direct treatment with ozone (dissolved in the physiological solution) and to ozonation byproducts (ozonated oils). Ozone's administration route influenced the inflammation inhibition, ozonated oils being the best anti-inflammatory agents. We found that the ozone dosage, meaning, the ozonation degree of oils, as well as the frequency of application, is a key factor in the biological effect of ozone-based therapies.

Acknowledgements

The authors thank the Department of Graduate Study and Investigation of the Instituto Politécnico Nacional of Mexico (Projects 20170481, 20170590) and the Consejo Nacional de Ciencia y Tecnología of Mexico—CONACyT (Projects 83593, 83275, 153356, 156150). P. Guerra-Blanco thanks CONACyT for the scholarship support. Some of this study was carried out as part of the activities of the "Laboratorio Nacional de Investigación y Desarrollo de Radiofármacos LANIDER-CONACyT.

Conflict of interest

Authors declare no conflict of interest.

Author details

Tatyana Poznyak[1]*, Pamela Guerra Blanco[1], Arizbeth Pérez Martínez[2], Isaac Chairez Oria[3] and Clara-L. Santos Cuevas[4]

*Address all correspondence to: tpoznyak@ipn.mx

1 Facultad de Ciencias Químicas, Universidad Autónoma de Chihuahua, Chihuahua, Mexico

2 Instituto Politécnico Nacional-ESIQIE, Mexico City, Mexico

3 Instituto Politécnico Nacional-UPIBI, Mexico City, Mexico

4 Instituto Nacional de Investigaciones Nucleares, Mexico City, Mexico

References

[1] Bocci V. Ozone, A New Medical Drug. Siena: Springer; 2005. pp. 1-3

[2] Altman N. Ozone: Life–Threatening Pollutant or Powerful Healing Agent? FAST, Bio-Oxidative Therapies: Oxygen, Ozone and H_2O_2; 1990. Editorial Healing Arts Press. pp. 3-12

[3] Bocci V. Scientific and medical aspects of ozone therapy: State of the art. Archives of Medical Research. 2006;**37**:425-435

[4] William PA, Kendall HN, Christopher FS, Jon FM, Louis IJ, Giuseppe SL, Davies JA. Free radical biology and medicine: It's a gas, man!! American Journal of Physiology–Regulatory, Integrative and Comparative Physiology. 2006;**291**(3):R491-R511

[5] Shoemaker JM. Ozone therapy: History, physiology, indications, results. Nottingham; 2005. Available from: http://www.fullcircleequine.com/oz_therapy.pdf

[6] Hunt NK, Marinas BJ. Kinetics of *Escherichia coli* inactivation with ozone. Water Research. 1997;**31**:1355-1362

[7] Knak JSJ. Oxidative stress and free radicals. Journal of Molecular Structure. 2003;**666**: 387-392

[8] Hernandez FA. To what extent does ozone therapy need a real biochemical control system? Assessment and importance of oxidative stress. Archives of Medical Research. 2007;**38**:571-578

[9] Madrid Declaration on Ozone Therapy. International Meeting of Ozone Therapy Schools. AEPROMO. Madrid, España: Spanish Association of Medical Professionals in Ozone Therapy (AEPROMO); 2010

[10] Pryor WA, Godber SS. Noninvasive measures of oxidative stress status in humans. Free Radical Biology & Medicine. 1991;**10**:177-184. DOI: 10.1016/0891-5849(91)90073-C

[11] Sergeev V, Burkin I, Levin A, Poznyak T. URSS Patent 1500943; 1989

[12] Razumovskii SD, Zaikov GE. Kinetics and mechanism of the reaction of ozone with double bonds. Russian Chemical Reviews. 1980;**49**:2344-2376

[13] Razumovskii SD, Zaikov GE. Ozone and Its Reactions with Organic Compounds. Amsterdam: Elsevier; 1984

[14] Poznyak TI, Kiseleva EV, Turkina TI. Distribution of the total unsaturation in lipid components of plasma as a new differential diagnostic method in clinical analysis. Journal of Chromatography A. 1997;**777**:47-50. DOI: 10.1016/S0021-9673(97)00559-1

[15] Poznyak T, García A, Kiseleva E. Ozone application for the people health state monitoring by the total unsaturation index determination–medigraphic.com. Rev Ozonoterapia. 2008;**1**:15-23

[16] Guerra-Blanco P, Poznyak T, Chairez I, Brito-Arias M. Correlation of structural characterization and viscosity measurements with total unsaturation: An effective method for controlling ozonation in the preparation of ozonated grape seed and sunflower oils. European Journal of Lipid Science and Technology. 2015;**117**:988-998. DOI: 10.1002/ejlt.201400292

[17] Criegee R. Mechanismus der Ozonolyse. Angewandte Chemie. 1975;**87**:765-771. DOI: 10.1002/ange.19750872104

[18] Bailey PS, editor. Ozonation in Organic Chemistry. 1st ed. Vol. 1. New York: Academic Press; 1978

[19] Sologub VK, Oliunina NA, Lisitsyn DM, Lavrov VA, Babskaia IE. Determination of double bonds in the lipid fraction of blood plasma in burn patients using the ADS-4M apparatus. Biulleten' Eksperimental'noĭ Biologii i Meditsiny. 1985;**100**:308-310

[20] Pérez A, Cuevas CLS, Chairez I, Poznyak T, Ordaz-Rosado D, García-Becerra R, et al. Ozone dosage effect on C6 cell growth, in vitro and in vivo tests: Double bond index for characterization. Analytical Methods. 2014;**6**:4567-4575. DOI: 10.1039/C4AY00104D

[21] Poznyak T, Puga JM, Kiseleva E, Martinez L. Chromium toxic effect monitoring using ozonation method. International Journal of Toxicology. 2002;**21**:211-217. DOI: 10.1080/10915810290096342

[22] Guerra-Blanco P, Poznyak T, Pérez A, Gómez y Gómez YM, Bautista-Ramírez ME, Chairez I. Ozonation degree of vegetable oils as the factor of their anti-inflammatory and wound-healing effectiveness. Ozone Science and Engineering. 2017;**39**:374-384. DOI: 10.1080/01919512.2017.1335185

[23] Castillo R, López A. Algunas consideraciones y experiencias sobre el uso del ratón atímico como modelo experimental. Interferón y Biotecnología. 1988;**5**(3):291-296

[24] Grobben B, De Deyn PP, Selegers H. Rat C6 glioma as experimental model system for the study of glioblastoma growth and invasion. Cell and Tissue Research. 2002;**310**:257-270

[25] Simos Y, Karkabounas S, Verginadis I, Charalampidis P, Filiou D, Charalapoulos K, Zioris I, Kalfakakou V, Evangellou A. Intra-peritoneal application of catechins and EGCG as in vivo inhibitors of ozone-induced oxidative stress. Phytomedicine. 2011;**18**(7):579-585

[26] Escobar A. Gliomas y angiogenesis. Revista Mexicana de Neurociencia. 2002;**3**:21-24

[27] Washüttl J, Viebahn R, Steiner I. The influence of ozone on tumor tissue in comparison with healthy tissue (in vitro). Ozone Science and Engineering. 1990;**12**:65-72. DOI: 10.1080/01919519008552455

[28] An Y-L, Nie F, Wang Z-Y, Zhang D-S. Preparation and characterization of realgar nanoparticles and theri inhibitory effect on rat glioma cells. International Journal of Nanomedicine. 2011;**6**:3187-3194

[29] Andreyev AY, Kushnareva YE, Starkov AA. Mitochondiral metabolism of reactive oxygen species. Biochemistry. 2005;**70**:200-214

[30] Schulz S, Häussler U, Mandic R, Heverhagen JT, Neubauer A, Dünne AA, Werner JA, Weihe E, Bette M. Treatment with ozone/oxygen-pneumoperitoneum results in complete remission of rabbit squamous cell carcinomas. International Journal of Cancer. 2008;**122**:2360-2367

[31] Vaupel P. Oxygenation of human tumors: evaluation of tissue oxygen distribution in breast cancers by computerized O_2 tension measurements. Cancer Research. 1991; **51**(12):3316-3322

[32] Vaupel P. Blood flow, oxygen consumption, and tissue oxygentaion of human breast cancer xenografts in nude rats. Cancer Research. 1987;**47**(13):3496-3503

[33] Garófano GJR, Venancio CG, Suazo CAT, Almeida PIF. Application of the wavelet image analysis technique to monitor cell concentration in bioprocesses. Brazilian Journal of Chemical Engineering. 2005;**22**(4):573-583

[34] Salganik IR. The benefits and hazards of antioxidants: Controlling appoptosis and other protective mechanisms in cancer patients and the human population. Journal of the American Collage of Nutrition. 2001;**20**:464S-472S

[35] Valacchi G, Fortino V, Bocci V. The dual action of ozone on the skin. The British Journal of Dermatology. 2005;**153**:1096-1100. DOI: 10.1111/j.1365-2133.2005.06939.x

[36] Travagli V, Zanardi I, Valacchi G, Bocci V. Ozone and ozonated oils in skin diseases: A review. Mediators of Inflammation. 2010;**2010**:9. DOI: 10.1155/2010/610418

[37] Sánchez GM, Re L, Perez-Davison G, Delaporte RH. Las aplicaciones médicas de los aceites ozonizados, actualización. Rev Esp Ozonoterapia. 2012;**2**:121-139

[38] Campanati A, De Blasio S, Giuliano A, Ganzetti G, Giuliodori K, Pecora T, et al. Topical ozonated oil versus hyaluronic gel for the treatment of partial- to full-thickness second-degree burns: A prospective, comparative, single-blind, non-randomised, controlled clinical trial. Burns Journal of the International Society for Burn Injuries. 2013;**39**:1178-1183. DOI: 10.1016/j.burns.2013.03.002

[39] Menéndez S, Falcón L, Maqueira Y. Therapeutic efficacy of topical OLEOZON® in patients suffering from onychomycosis. Mycoses. 2011;**54**:e272-e277. DOI: 10.1111/j.1439-0507.2010.01898.x

[40] Sadowska J, Johansson B, Johannessen E, Friman R, Broniarz-Press L, Rosenholm JB. Characterization of ozonated vegetable oils by spectroscopic and chromatographic methods. Chemistry and Physics of Lipids. 2008;**151**:85-91. DOI: 10.1016/j.chemphyslip.2007.10.004

[41] Díaz M, Lezcano I, Molerio J, Hernández F. Spectroscopic characterization of ozonides with biological activity. Ozone Science and Engineering. 2001;**23**:35-40. DOI: 10.1080/01919510108961986

[42] Díaz MF, Hernández R, Martínez G, Vidal G, Gómez M, Fernández H, et al. Comparative study of ozonized olive oil and ozonized sunflower oil. Journal of the Brazilian Chemical Society. 2006;**17**:403-407. DOI: 10.1590/S0103-50532006000200026

[43] Skalska K, Ledakowicz S, Perkowski J, Sencio B. Germicidal properties of ozonated sunflower oil. Ozone Science and Engineering. 2009;**31**:232-237. DOI: 10.1080/01919510902838669

[44] Kataoka H, Semma M, Sakazaki H, Nakamuro K, Yamamoto T, Hirota S, et al. Proinflammatory event of ozonized olive oil in mice. Ozone Science and Engineering. 2009;**31**:238-246. DOI: 10.1080/01919510902839022

[45] Valacchi G, Lim Y, Belmonte G, Miracco C, Zanardi I, Bocci V, et al. Ozonated sesame oil enhances cutaneous wound healing in SKH1 mice. Wound Repair and Regeneration. 2011;**19**:107-115. DOI: 10.1111/j.1524-475X.2010.00649.x

[46] Valacchi G, Zanardi I, Lim Y, Belmonte G, Miracco C, Sticozzi C, et al. Ozonated oils as functional dermatological matrices: Effects on the wound healing process using SKH1 mice. International Journal of Pharmaceutics. 2013;**458**:65-73. DOI: 10.1016/j.ijpharm.2013.09

Chapter 4

Ozone in Dentistry

Aysan Lektemur Alpan and Olcay Bakar

Additional information is available at the end of the chapter

http://dx.doi.org/10.5772/intechopen.75829

Abstract

Ozone (triatomic oxygen or trioxygen) is the combination of three naturally occurring oxygen atoms. Ozone therapy is an alternative to traditional approaches in dentistry. The main feature suggests that ozone can be used in dentistry as a strong antimicrobial agent. In addition, ozone has antimicrobial, immune system regulatory, metabolic rate, and biosynthesis-enhancing effects. Ozone affects cellular and humoral immunity. It has positive effects on oxygen transport in the body; production of adenosine triphosphate (ATP); and production of enzymes such as glutathione peroxidase, catalase, and superoxide dismutase. Ozone use in dentistry can be made possible via ozone gas, ozonated water, ozonized olive, or sunflower oil. Ozone is used in periodontology (gingivitis, periodontitis, periimplantitis, surgical injuries, prophylaxis), oral pathologies (stomatitis, aphthous ulceration, candidiasis, herpes infections), endodontics (root canal treatment, the fistula, abscesses), oral surgery (hemostasis, wound healing, implantation, reimplantation, tooth extraction), prosthodontics (disinfection of crowns, disinfection of the alloy part of partial dentures), orthodontics (TME function disorders, trismus, myoarthropathies), and restorative dentistry (caries, dentine hypersensitivity, cracked tooth syndrome, bleaching, disinfection of cavity). As a result of the studies performed, ozone therapy in dentistry should be considered as an aid to conventional treatments.

Keywords: antioxidant, antimicrobial, dentistry, immunostimulating, ozone

1. Introduction

Ozone (triatomic oxygen or trioxygen) is the combination of three naturally occurring oxygen atoms. Ozone is present in the gas form in the concentration of 1–10 ppm in the stratosphere in nature. Molecular weight is 47.98 g/mol, and it is highly endothermic and also thermodynamically unstable as an oxygen compound. Depending on environmental

conditions, such as short half-life, pressure, and heat, molecular oxygen in ozone can be converted to atomic oxygen in a short time [1]. Ozone is the third most powerful oxidant known that does not possess radical properties due to its chemical structure [2]. Ozone has a higher energy than atmospheric oxygen, 1.6 times more dense, and 10 times more soluble in water than oxygen [1]. In 1785, Van Marum noticed that when the spark occurred in the electrostatic machine, there was a peculiar smell of air around him. In 1801, Cruickshank heard the same smell on the anode side during electrolysis of water. Sconbein described this substance in 1840 as "Ozein," meaning to smell it in Greek. In 1856, Werner Von Siemens designed an ozone generator in 1857, which was used in the disinfection of operating theaters. Since these types of generators are the forerunners of later generations, these types of generators in the market are called "Siemens type" ozone generators. In 1860, the first ozone generator in Monaco was used for plant treatment. In 1870, for the first time in medical treatment, it was used by Lender [3].

In the nineteenth century, Dr. Fisch used ozonated water in his practice for the first time in dentistry and introduced it to Dr. Erwin Payr who was a German surgeon. Dr. Erwin Payr used ozone in surgery and he reported a publication (of 290 pages) entitled "Ozone Treatment in Surgery" (Über Ozonbehandlung in der Chirurgie) at the 59th Meeting of the German Surgical Society in 1935 [4].

As modern science is used in the practice of dentistry, it is also changing and developing with time. Ozone therapy is an alternative to traditional approaches and can be considered as a model to help in healing. Emerging technology has led us to less invasive and more conservative work.

The main feature suggests that ozone can be used in dentistry as a strong antimicrobial agent. It is effective against Gram (+, −) bacteria, viruses, and fungi. Ozone, which is used in dental prosthesis, endodontics, restorative dentistry, periodontology, and oral and maxillofacial surgery, offers great advantages in addition to traditional treatments [5, 6].

Ozone shows the antimicrobial effect by creating cell membrane damage. It reacts with the double bonds of the hydrocarbons in the cell membrane and causes the modification of the cell content by the action of the secondary oxidant. Ozone is highly effective against antibiotic-resistant species. The antimicrobial activity of ozone is increased in liquid and acidic pH [6]. The basis of the ozone action mechanism in viral infections is inhibiting infected cell peroxide sensitivity and the synthesis of viral proteins by altering the activity of the reverse transcriptase enzyme [7].

Ozone affects cellular and humoral immunity. It stimulates the proliferation and immunoglobulin synthesis in immune cells, accelerates the sensitivity of macrophage phagocytosis, and activates other macrophage functions. This activation results in the production of specific molecules called cytokines. This suggests that ozone administration at low doses is beneficial to people with an impaired immune system. Ozone also promotes the synthesis of biologically active molecules such as interleukins, leukotrienes, and prostoglandins, thereby helping to reduce inflammation and improve wound healing [6, 8].

Ozone increases the partial oxygen pressure in the tissues and increases the oxygen transport in the body causing changes in cell metabolism. This change increases the use of oxygenated respiration and therefore energy sources (glycolysis, cycling of the Krebs, β-oxidation of fatty acids). It also prevents the erythrocytes from collapsing and increases the contact surface of erythrocytes for oxygen transport. It activates the Krebs cycle, which stimulates adenosine triphosphate (ATP) production, and causes a significant reduction in nicotinamide adenine dinucleotide (NADH) leading to oxidation of cytochrome C. Ozone therapy stimulates production of enzymes such as glutathione peroxidase, catalase, and superoxide dismutase which act as free radical scavengers [9].

It promotes intracellular protein synthesis by stimulating mitochondria and ribosomes. This alteration may lead to activation of cell functions and regeneration potential of tissues and organs. Ozone causes dilation in arterioles and venules by stimulating the release of vasodilators such as nitric oxide [6, 10].

1.1. Dental treatment modalities of ozone therapy

- Biofilm purging (elimination of bacterial pathogens) [5].
- Periodontal pocket disinfection and osseous disinfection.
- Prevention of dental caries.
- Endodontic treatment.
- Tooth extraction.
- Tooth sensitivity.
- Temporomandibular joint treatment.
- Gingival recession (exposed root surfaces).
- Pain control.
- Infection control.
- Accelerating of wound healing.
- Tissue regeneration.
- Controlling halitosis.
- Remineralization of tooth surface.
- Teeth whitening (bleaching).

Ozone use in dentistry can be possible via ozone gas, ozonated water, ozonized olive, or sunflower oil [2]. Ozone, a very unstable molecule in gas form, lasts a few minutes in the air, while the aquatic life lasts a few days. However, it has been reported that ozone can be measured for months and years when dissolved in an oil-based content such as 100% pure olive oil [11].

1.2. Ozone toxicity

It should not be forgotten that ozone is a toxic gas if it is inhaled. Eyes and lungs are very susceptible to ozone. For this reason, long-term exposure to ozone results in some side effects, such as epiphora, irritation of the upper airways, bronchoconstriction, rhinitis, cough, headache, and vomiting may occur, depending on the time of the ozone exposure. In such cases, it has been reported that administering supportive treatment such as oxygen, ascorbic acid, vitamin E, and N-acetylcysteine to the patient would be beneficial [12–14]. Pathological and anatomical studies showed that blood clotting is impaired in a typical table ozone poisoning and has been shown to occur in lung hematomas. However, the ozone gas is not a chemical disinfectant, and after completing the disinfection task due to its unstable structure, it rapidly transforms into oxygen [2].

As ozone therapy is an atraumatic treatment method, the areas of use in dentistry are increasing with the protection of healthy and decayed dental caries, disinfection of dental unit water systems, antibacterial effect in avulse teeth, and healing properties of oral lesions.

1.3. Contraindications of ozone therapy

- Pregnancy [7].
- Hyperthyroidism.
- Glucose-6-phosphate-dehydrogenase deficiency.
- Severe anemia.
- Severe myastenia.
- Active hemorrhage.
- Acute alcohol intoxication.

Ozone can be used for prophylaxis in dentistry due to its biological properties and for the treatment of various diseases. Ozone is used in periodontology (gingivitis, periodontitis, periimplantitis, surgical injuries, prophylaxis), oral pathologies (stomatitis, aphthous ulceration, candidiasis, herpes infections), endodontics (root canal treatment, the fistula, abscesses), surgical procedures (hemostasis, wound healing, implantation, reimplantation, tooth extraction), prosthodontics (disinfection of crowns, disinfection of the alloy part of partial dentures), orthodontics (TME function disorders, trismus, myoarthropathies), and restorative dentistry (caries, dentine hypersensitivity, cracked tooth syndrome, bleaching, disinfection of cavity) [15].

1.4. Ozone gas generating systems

1. Corona discharge ozone generators: with the corona discharge method, ozone gas (O_3) is formed by breaking the double bond of the oxygen molecule (O_2) by passing electric current and combining the other free oxygen atom [6].

http://dx.doi.org/10.5772/intechopen.75829

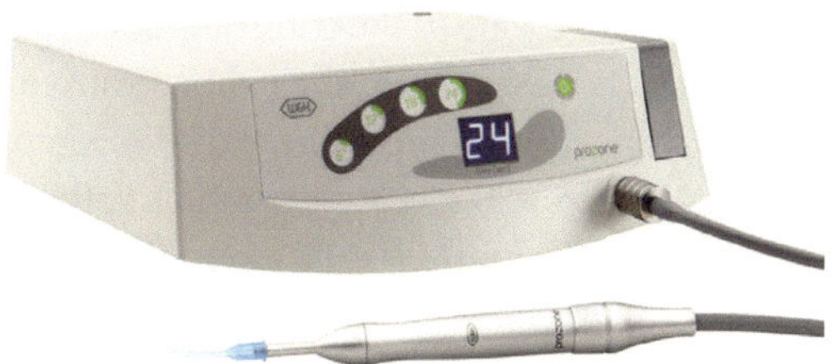

Figure 1. An ozone-generating system used in dental practice [17].

2. Ultraviolet ozone generators: the ultraviolet method is used to break up the oxygen molecule by passing it through an ultraviolet bulb which emits light with a wavelength shorter than 220 nm, and the released oxygen atom combines with the other oxygen molecule to form the ozone gas.

3. Cold plasma system: it is used for purification of air and water.

Thanks to innovative technology, ozone has become painless, safe, effective, and easy to apply in many areas of medicine. New-generation medical ozone generators can produce ozone at very narrow therapeutic ranges (0.1–2.1 μg /s) from oxygen molecules present in the atmosphere or in liquids. The applied microcurrent (max 100 μA) is completely harmless to both the patient and the implementer [16] (**Figure 1**).

2. Role of ozone on restorative dentistry

In recent years, ozone treatment has begun to be used as a new method in the treatment of caries. It has been suggested that application of ozone to caries stops or hardens these lesions. Application of ozone to caries will provide an alternative to conventional treatment modalities. It was demonstrated in studies that ozone can be used to eradicate bacteria in carious lesions, painlessly.

Baysan et al. [18], found a significant decrease in *Streptococcus mutans* and *Streptococcus sobrinus* numbers on primary root carious lesions which were applied ozone gas for a period of 10 s. Then, the in vitro study was adapted to a randomized clinical trial and the results of the controls were measured by using DIAGNOdent and ECM. A significant increase in remineralization was observed in ozone groups. In a randomized trial, Holmes evaluated the effect of ozone on surface hardness (soft, brittle, stiff) on root caries. At the end of 12 months, 100% of the teeth treated with ozone treatment had hardened caries surfaces, and 37% of caries in the control group without treatment reported that the lesions were getting worse [19]. Samuel et al [20] evaluated the effect of ozonated water in remineralizing artificially created initial enamel caries using laser fluorescence and polarized light microscopy. According the results, reduced DIAGNOdent scores and greater depth of remineralization were gained following application of ozonated water and the ozone-treated group exhibited maximum remineralization under the polarized light microscopy. Polydorou et al. [21] used two different bonding systems, 40 and 80 s, in ozone-applied cavities in an in vitro study. Two different bonding systems were performed without any application of ozone to the control group. The teeth were then restored with composite resin. *Streptococcus mutans* ratio in the ozone group for 80 s values was statistically decreased in comparison to the other groups. These results are promising for applications directed at *Streptococcus mutans,* which is the most important pathogen responsible for caries. On the other hand, in a Cochrane review, authors concluded that application of ozone on caries provides no evidence in terms of arresting or reversing the decay process [22].

Tooth structure can lessen via attrition, abrasion, erosion, trauma from occlusion, and it may cause wearing away of enamel and dentin, thereby causing hypersensitivity. Ozone application has been found to reduce sensitivity in exposed enamel and dentin and also in cases of root sensitivity. It was found that 40–60 s application of ozone offers pain reduction in sensitive teeth. Ozone initiates removal of the smear layer, opens the dentinal tubules, and widens them so that the remineralizing agents—calcium and fluoride ions—can enter the dentinal tubules easily, readily, and completely, preventing the fluid exchange from dentine tubules. Depending on this, termination of sensitivity occurs following ozone application in a short time and also lasts longer than conventional treatment modalities [9].

Delay and Holmes [23] reported that the ozone application provides a reduction in the symptoms of patients with cracked tooth syndrome. Medozon, an ozone-generating device, claims that ozone application of 60–120 s to the cracked area in the fractured tooth syndrome provides long-lived restorative material [24].

Ozone can be used on root canal-treated discolored teeth by irradiating the root canal for 3 min. This treatment provides good esthetic result by bleaching the tooth. Tessier et al. [25] evaluated the ozone efficacy in an experimental rat model used to lighten tetracycline-stained incisors. At the end of the study, it was found that ozone application could be successfully used for lightening the yellowish tinge of tetracycline-stains incisors.

3. Role of ozone on periodontology

The main ozone application area in periodontology is relayed on its antimicrobial properties. It seems to be effective against both Gram (+) and Gram (−) bacteria, viruses, and fungi. It can be applied into the periodontal pocket with the different tips of generators (**Figure 2**), ozonated water, or ozonated oil.

Nagayoshi et al. [27] investigated the effect of ozonated water on cell permeability and viability of microorganisms. Gram-negative bacteria (*Porphyromonas gingivalis, Porphyromonas endotalis*) was found to be more sensitive to ozone than streptococci and *Candida albicans*. In addition, the ozonated water has strong bactericidal action against *Streptococcus mutans* bacteria in the plaque biofilm. Also it was reported that ozonated water inhibited the experimental bacterial plaques in vitro. In another study, it was concluded that high concentrations of ozone water (20 $\mu g\ ml^{-1}$) had an antibacterial effect equal to 0.2% concentration of chlorhexidine, whereas highly concentrated ozone gas ($\geq 4\ g\ m^{-3}$) had an antibacterial effect of as much as 2% chlorhexidine and more effective than 0.2% chlorhexidine [28]. Ramzy et al. [29] used 150 ml of ozonized water for periodontal pocket irrigation (5–10 min once weekly, 4 weeks)

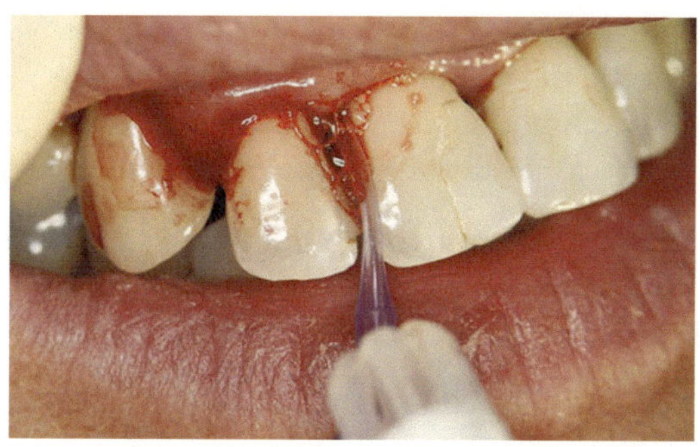

Figure 2. Ozone application into periodontal pocket [26].

in patients suffering from aggressive periodontitis. Statistically significant decreases in terms of pocket depth, plaque index, gingival index, and bacterial count were observed. In a study, authors compared the effect of oral irrigation with ozone water, 0.2% chlorhexidine and 10% povidone iodine, in chronic periodontitis patients and concluded that local ozone application could be used as a powerful atraumatic and antimicrobial agent in the nonsurgical treatment of periodontal disease for both home care and professional practice [30].

In contrast, Eltaş and Yavuzer applied ozone gas in addition to scaling and root planning in acute gingivitis patients. Changes in plaque index, pocket depth, and clinical attachment levels did not differ significantly between groups at 4 weeks after treatment [31]. Yılmaz et al. [32] investigated the changes in clinical and microbiological parameters of mechanical treatment, mechanical treatment + erbium: yttrium-aluminum-garnet laser, and mechanical treatment + gas ozone application in chronic periodontitis. Attachment gain and pocket depth reduction were found to be greater in the laser group than in the other groups. Although not statistically significant, the decrease in anaerobic flora was observed in both laser and ozone groups.

Karapetian et al. compared ozone therapy with surgical procedures and conventional methods in patients with periimplantitis and stated that the most effective method to eliminate bacteria was ozone therapy [33]. In an in vitro study, gaseous ozone (140 ppm, 33 mL/s) for 6 and 24 s was applied to saliva-coated titanium (SLA and polished) and zirconia (acid etched and polished) disks to determine the antibacterial effect on periimplantitis caused by bacteria such as *Streptococcus sanguinis* and *Porphyromonas gingivalis*. Gaseous ozone showed selective efficacy to reduce adherent bacteria on titanium and zirconia without affecting adhesion and proliferation of osteoblastic cells [34].

4. Role in bone regeneration

Besides the antiseptic and disinfectant properties of ozone, it was also investigated for its effects on bone regeneration in recent years. One of the first study about this subject belonged to Özdemir et al. [35]. According this study results, ozone application combined with autograft provided an increase the amount of total bone area and osteoblast count.

Kazancioglu et al. [36] compared the effects of low-dose laser therapy and ozone treatment on bone regeneration in 5-mm critical-size defects in rats, and all defects were restored with biphasic calcium phosphate grafts. According to the histomorphometric measurements, the new bone area in the ozone group was statistically higher than the control and low-dose laser group.

In another study comparing the effects of hyperbaric oxygen and systemic ozone administration, the rats were sacrificed on days 5, 15, and 30, postoperatively. There was no difference in bone formation between hyperbaric oxygen and ozone [37].

Lektemur Alpan et al. [38] used diabetic rat calvarial defects with xenograft, and they concluded that the ozone accelerated bone morphogenetic protein-2 and osteocalcin positivity followed by accelerated xenograft resorption and enhanced bone regeneration.

5. Role of ozone in oral surgery

Ozone therapy has a vast range of applications in oral surgery because of its biological properties such as enhancing wound healing, improving several properties of erythrocytes, and facilitating oxygen release in the tissues. All these biological events cause and hence improve the blood supply to the ischemic zones leading to use of ozone in cases of wound-healing impairments, following surgical interventions like tooth extractions or implant dentistry.

Ozone treatment can be applied in cases such as disinfection of wound area, treatment of soft tissue lesions (aphthous ulcers, herpes simplex, herpes zoster, etc.), healing disorders in bone and soft tissue, alveolitis, periimplantitis, bisphosphonate-related osteonecrosis, tooth transplantation, and decontamination of root surfaces of avulsed teeth planned to be reimplanted [39]. It is possible to apply ozonized water to infections that may occur after osteotomy in oral surgery. In some prospective studies, it has been shown clinically and histologically that ozonized water has a positive effect on soft tissue healing. In a prospective study involving 250 patients, application of ozone water as a cooling and flushing agent during third molar osteotomies has been shown to reduce infectious complications after surgery [39]. Kazancioglu et al. evaluated the effect of ozone therapy on pain, swelling, and trismus following third molar surgery, and they concluded that ozone application effectively reduced postoperative pain; however, it had no effect on swelling and trismus [40].

Ahmedi et al. [41] evaluated the efficacy of ozone gas on the reduction of dry socket, which occurred after surgical extraction of lower jaw third molars. Two groups were evaluated: in the control group, saline solution was used for irrigation of extraction sockets and, intra-alveolar ozone was applied at 12 s (Prozone, W&H, UK) in the experimental group. They concluded that the ozone gas has a positive effect on reducing the development of dry socket and pain following third molar surgery depending on metabolic capabilities of ozone for promoting hemostasis, increasing the supply of oxygen, and inhibiting bacterial proliferation.

In a study, ozone therapy was compared with the photo-biomodulation therapy in mental nerve injury by counting Schwann cells and fasciculated nerve branches and measuring fascicular nerve areas. At the end of the study, a better healing pattern was observed in the treatment groups. The number of Schwann cells was markedly larger in the ozone treatment and photo-biomodulation groups than in the control group [42].

The effects of ozone therapy stimulating cell proliferation and soft tissue healing must be taken into account in the treatment of bone necrosis in patients using bisphosphonates [43]. Many studies have been carried out on the use of ozone in osteonecrosis cases occurring in jaws due to the use of bisphosphonates. Similar results were obtained in these clinical trials with different routes of administration of ozone (gas, ozonated water, and oil) [44–46]. After radiotherapy in maxilla or mandible, the amount of oxygen in the affected area is considerably reduced. Radiotherapy leads to obliteration of intrabony vessels and inadequate vascular support in spongiosal medullary spaces resulting in xerostomia, mucositis, or loss of taste sensation. As a result, fibrosis and aseptic osteonecrosis may occur. Recovery after surgical procedures is impaired after tooth extraction from this kind of bone in comparison to healthy bones which have adequate blood supply. Such cases are always at risk of persistent osteoradionecrosis [47].

Akdeniz et al. [48] performed a study on human primary gingival fibroblasts exposed to cytotoxic concentrations of bisphosphonates. They concluded that ozone gas plasma therapy significantly decreased the genotoxic damage and this application provided 25%, 29%, and 27% less genotoxic damage, respectively, in bisphosphonate groups and improved the wound closure rate on human gingival fibroblasts.

Doğan et al. [49] investigated the effects of ozone on cancer progression and survival with radiotherapy. Experimental tongue cancer was formed in rats and were separated into four groups. Ozone groups were received 1 ml at a concentration of 15 mcg/ml ozone (rectal 4 sessions, for 5 days after 22 week). Groups that received ozone showed more histopathologic improvements in comparison to other groups. Radiotherapy combined with ozone therapy has provided more survival rate and tumoral reduction than the other groups.

6. Role in prosthodontics

Dentures are commonly inhabited by microbial plaque, especially *Candida albicans*. Denture stomatitis is routinely encountered in clinical practice, which can be prevented by effective denture plaque control. One successful method to do so is the use of ozone as disinfecting agent to clean denture. Arita et al. [50] concluded that exposure of dentures to flowing ozonated water (2 or 4 mg/l) for 1 min can reduce the number of *Candida albicans*.

Oizumi el al. [51] compared the microbicidal effect of gaseous ozone with that of ozonated water on oral microorganisms (*Streptococcus mutans, Staphylococcus aureus, Candida albicans*). They concluded that direct exposure to gaseous ozone seems to be a more effective microbicide than ozonated water for decreasing the microorganisms.

In another study, ozonized olive oil efficacy was evaluated in the treatment of oral lesions and conditions (aphthous ulcerations, herpes labialis, oral candidiasis, oral lichen planus, and angular cheilitis) (**Figure 3**). The ozonized olive oil was applied twice. All of the conditions showed improvement in the signs and symptoms at the end of 6 months [52].

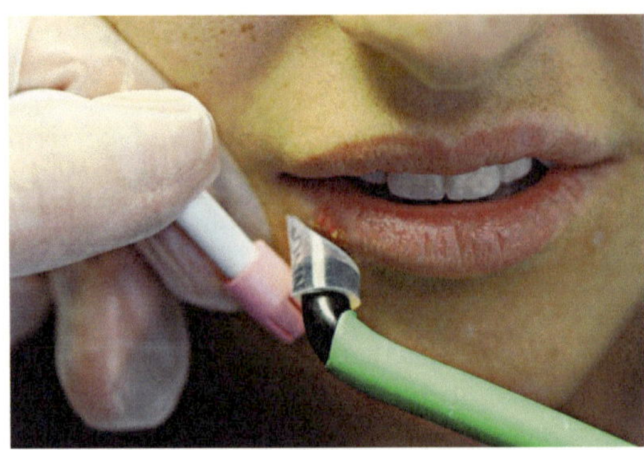

Figure 3. Ozone application on herpes labialis.

Temporomandibular disorder (TMD) is a pathological condition involving both the muscular and skeletal system in the temporomandibular joint region (TMJ). This is characterized by pain in the preauricular region during jaw movements, limitation during mandibular movements, pain during chewing muscles and palpation in the TMJ region, and TMJ voices. Pain usually occurs during chewing or mandibular movements [53]. Temporomandibular disorder is a collective term embracing several problems that involve the temporomandibular joint, masticatory muscles, or both and treated usually with conservative and reversible therapy. Regular application of the ozone to the TMJ region with special probes developed for deep tissue stimulation allows for access to deep tissue under the skin. Ozone application increases the oxygenation of muscle and cartilage tissue and the creation of anti-inflammatory effect. This can be used as a noninvasive treatment method in patients with TMD. In addition, there are different current treatment modalities reported in the literature, including medication therapy, low-level laser therapy, vibratory stimulation, and, more recently, bio-oxidative ozone therapy which reduced pain in the TMJ region and improved in TMD-induced mouth opening problems after regular treatments [54, 55].

7. Role of ozone in endodontics

Ozone is intensively used in root canal therapy due to its strong antimicrobial properties and absence of cytotoxicity. Ozone can be an effective agent when it is used in adequate concentration, time, and applied in a correct way into the root canals after other treatment steps have been performed. Most of the studies on effect of ozone in endodontics investigate its antimicrobial activities in the form of ozone gas, ozonated water, and ozonized oil.

Ozone is a powerful antibacterial agent. In a study, ozone was found to disinfect the bovine tooth dentin tubules effectively [56]. Nagayoshi et al. [27] demonstrated that in concentrations of 0.5–4 mg/L, ozonated water killed pure cultures of *Porphyromonas endodontalis* and *Porphyromonas gingivalis* effectively. These species were found more vulnerable to ozonated water than Gram-positive oral *Streptococci* and *Candida albicans*.

In a study, Hems et al. [57] evaluated antibacterial potential of gas form (produced by Pure zone device) and aqueous (optimal concentration 0.68 mg/L) ozone on the test species

http://dx.doi.org/10.5772/intechopen.75829

Enterococcus faecalis. They found that ozone in solution has antibacterial effect on planktonic *Enterococcus faecalis* after 240 s treatment; however, it shows not much antibacterial effect on *Enterococcus faecalis* within a biofilm.

Estrela et al. [58] studied two forms of ozone—ozonated water and gaseous ozone—with 2.5% hypochlorite and 2% chloehexidine in infected dental root canals. All agents contacted 20 min and none of them had killed *Enterococcus faecalis* in human-infected root canals.

Use of gas delivery of ozone at a flow rate of 0.5–1 1/min with a net volume of 5 gm/ml for 2–3 min gave favorable result in eliminating pathogen species in the root canal [59].

As an intracanal irrigant, ozonated water can be used in infected necrotic canals, and as intracanal dressing, ozonized oils can be used to reduce target anaerobic biota. Ozone also enhances tissue regeneration and bone healing when used as a canal irrigant. Moreover, ozone water with sonification has antimicrobial effect in comparison to 2.5% NaOCl when used in the disinfection of the root canal [60].

Conflicting reports in term of antimicrobial effect of ozone on endodontic infections were presented in a review [61]. As a result, contradictory results regarding the efficacy of endodontic ozone administration have been reported in the literature.

8. Role of ozone in orthodontics

In orthodontic treatments, diffuse opacity of enamel is commonly seen due to the effect of bonding material on enamel surface, as well as white spot lesions have been seen in first 4 weeks of the treatment. White spot lesion formation usually begins at bracket and toot interface and can reach beneath the bracket area. Hence, prophylactic therapy of enamel has an immense importance in orthodontic treatments.

Ghobashy et al. [62] studied on reducing demineralization of enamel bonded to the orthodontic bracket using ozonized olive oil. Patients who used ozonized olive oil gel with traditional oral hygiene instructions had significantly less decalcification areas during the orthodontic treatment.

Ozone also has a strong oxidizing effect that might cause weak adhesions between tooth and resin due to the negative effect of oxygen inhibition of polymerization. Cehreli et al. [63] evaluated the effect of prophylactic ozone pretreatment of enamel on shear bond strength of orthodontic brackets bonded with total or self-etch adhesive systems. Study revealed that ozone pretreatment of enamel did not have an effect on the shear bond strength of adhesive systems. Shear bond strength values of specimens in ozone group were even slightly higher.

9. Role of ozone in pedodontics

Ozone treatment has become more and more popular in the dental clinic every day, and it has become effective in many treatment and application procedures in pedodontics.

Applications of ozone in pediatric dentistry [4].

- **a.** Application in initial caries (remineralization effect).
- **b.** Application in root-surface caries (antibacterial effect).
- **c.** Use as a root channel disinfectant (antibacterial effect).
- **d.** Accelerate wound healing after surgical treatment.
- **e.** Treatment of oral ulcers and aphthae.
- **f.** Treatment of temporomandibular joint dysfunctions and irregularities.
- **g.** In tooth bleaching applications.

The use of ozone has positive effects on children, especially in terms of cooperation, such as not making noise, having very small hoods, not generating heat or bad smell, water spray or loud sounds of suctions, and not needing hand tools [4, 64]. Ozone prevents the caries formation via inhibiting the reproduction of pathogenic microorganisms, or destroying the cell wall by neutralizing or blocking [21, 65, 66] (**Figure 4**).

During this time, ozone attacks glycoproteins, glycolipids, and other amino acids and blocks enzymatic control systems of cells. Thus, the permeability of the cell membrane increases, extending to stop cell viability. After that, the ozone molecules can quickly enter the cell and cause the death of microorganisms [66].

In addition, ozone is an oxidative agent and can provide remineralization of demineralized dentin [21, 65]. The strongest acid naturally produced by acidogenic bacteria during caries formation is pyruvic acid. Pyruvic acid reacts with ozone and decarboxylates oxidatively to acetate and carbon dioxide. The remineralization of the initial caries lesions is supported by the buffered plaque fluid formed by the production of acetate [68]. Theoretically, ozone can be

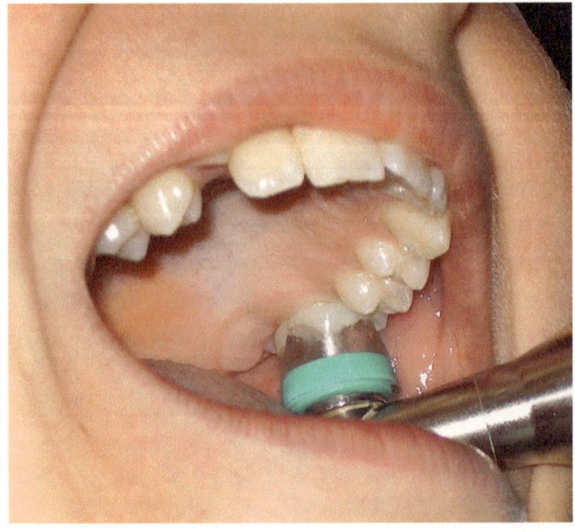

Figure 4. Ozone can be used easily on children [67].

used to reduce the number of bacteria in active caries lesions and consequently can temporarily stop caries progression; caries restoration can be delayed or prevented [22, 69].

In particular, studies have shown that ozone efficacy on pit and fissures where bacterial elimination was very difficult and the most susceptible areas for development of caries [70].

In a randomized clinical trial conducted by Huth et al. [71], 41 children aged 3–7 years were evaluated. A total of 51 patients with pairs of teeth having cavitation-free initial decay were separated into two groups and 40 s ozone (HealOzone-Kavo Dental GmbH Germany) was administered to the study group. After 3 months of clinical observation and DIAGNOdent measurements, regression and remineralization of initial caries were observed in ozone-treated teeth, but the results were not statistically significant.

In dental ozone applications, it is aimed to fix early lesions without altering the anatomical shape of the tooth, thanks to specially developed prophylactic tips. The first patient group to use this technique, which prevents unnecessary hard tissue loss in anterior and posterior initial caries lesions, is children.

A study was conducted with apprehensive children to determine ozone efficacy in open single-surface caries lesions. A total of 82 patients with single-surface caries lesions were separated into two groups and ozone was not applied to the control group, while ozone (HealOzone-Cavo Dental GmbH Germany) was applied to the other group for 20 sec. In ozone-treated group, hardness values improved in comparison to the control group. At the same time, when cooperative evaluation and assessment of dental anxiety was conducted on children, it was stated that ozone application was less worrying in child patients and more acceptable due to short-time application, sound, and water withdrawal [72].

In cavitated lesions, especially ozone gas application provides an antibacterial effect in the cavity surface and satisfies intended to stop progression of the lesion. Although there is not enough clinical experience about the interval and dose of ozone gas to prevent the lesion becoming active after a certain period of cavitated caries lesions, it is thought that promising results can be obtained especially in children who have difficulties in cooperativeness, and this technique may be developed widely and used in routine clinical practice.

10. Conclusion

Scientific studies show that ozone can be a promising therapeutic agent in the practice of dentistry. Besides atraumatic application and antimicrobial effects of ozone, carrying toxic risk and can have deadly consequences of wrong actions also incomplete understanding of the mechanism of action many clinicians approach suspicious to ozone. Considering the studies done so far, it can be said that ozone can be used as an additional application besides applications such as antiseptics and local antibiotics, which are given in addition to dental treatments. As a result, more clinical studies on ozone therapy should be performed and well-defined parameters should be established. However, more studies on ozone are required in order to be used routinely in dental treatments.

Author details

Aysan Lektemur Alpan[1*] and Olcay Bakar[2]

*Address all correspondence to: ysnlpn@gmail.com

1 Faculty of Dentistry, Department of Periodontology, Pamukkale University, Denizli, Turkey

2 Faculty of Dentistry, Department of Periodontology, Erzincan University, Erzincan, Turkey

References

[1] Bocci V. How ozone acts and how it exerts therapeutic effects. In: Lynch E, editor. Ozone: The Revolution in Dentistry. London: Quintessence Publishing Co; 2004. pp. 15-22. ISBN 1-85097-088-2

[2] Bocci VA. Scientific and medical aspects of ozone therapy. State of the art. Archives of Medical Research. 2006;**37**(4):425-435. DOI: 10.1016/j.arcmed.2005.08.006

[3] Makkar S, Makkar M. Ozone-treating dental infections. Indian Journal of Stomatology. 2011;**2**(4):256-259. Avaible from: http://indianjournalofstomatology.com/files/dec/11.pdf [Accessed December 29, 2017]

[4] Azarpazhooh A, Limeback H. The application of ozone in dentistry: A systematic review of literature. Journal of Dentistry. 2008;**36**(2):104-116. DOI: 10.1016/j.jdent.2007.11.008

[5] Saini R. Ozone therapy in dentistry: A strategic review. Journal of Natural Science, Biology and Medicine. 2011;**2**(2):151-153. DOI: 10.4103/0976-9668.92318

[6] Bhateja S. The miraculous healing theraphy; ozone theraphy in dentistry. Indian Journal of Dentistry. 2012;**3**:150-155. DOI: 10.1016/j.ijd.2012.04.004

[7] Bocci V. Ozone: A New Medical Drug. 2nd ed. Netherlands: Springer Publishing; 2005,2010. pp. 5-234. DOI: 10.1007/978-90-481-9234-2

[8] Celiberti P, Pazera P, Lussi A. The impact of ozone treatment on enamel physical properties. American Journal of Dentistry. 2006;**19**(1):67-72. PMID: 16555661

[9] Tiwari S, Avinash A, Katiyar S, Iyer A, Jain S. Dental applications of ozone therapy: A review of literature. The Saudi Journal for Dental Research. 2017;**8**(1-2):105-111. DOI: 10.1016/j.sjdr.2016.06.005

[10] Das S. Application of ozone therapy in dentistry. Indian Journal of Dental Advancements. 2011;**3**:538-542. ISSN: 2229-5038

[11] Mosallam RS, Nemat A, El-Hoshy A, Suzuki S. Effect of oleozon on healing of exposed pulp tissues. Journal of American Science. 2011;**7**:34-44. ISSN: 1545-1003

[12] Bocci V. Is it true that ozone is always toxic? The end of a dogma. Toxicology and Applied Pharmacology. 2006;**216**(3):493-504. DOI: 10.1016/j.taap.2006.06.009

[13] Bocci V, Aldinucci C. Biochemical modifications induced in human blood by oxygenation-ozonation. Journal of Biochemical and Molecular Toxicology. 2006;**20**(3):133-138. DOI: 10.1002/jbt.20124

[14] Nogales CG, Ferrari PH, Kantorovich EO, Lage-Marques JL. Ozone therapy in medicine and dentistry. The Journal of Contemporary Dental Practice. 2008;**9**(4):75-84. PMID:18473030

[15] Gupta G, Mansi B. Ozone therapy in periodontics. Journal of Medicine and Life. 2012; **5**(1):59-67. PMCID: PMC3307081

[16] Cardoso CC, Carvalho JC, Ovando EC, Macedo SB, Dall'Aglio R, Ferreira LR. Action of ozonized water in preclinical inflammatory models. Pharmacological Research. 2000;**42**(1):51-54. DOI: 10.1006/phrs.1999.0646

[17] Avaible from: http://www.wh.com/en_global/dental-products/prophylaxis-periodontology/ozonedevices/prozone/ [Accessed: December 30, 2017]

[18] Baysan A, Whiley RA, Lynch E. Antimicrobial effect of a novel ozone- generating device on micro-organisms associated with primary root carious lesions in vitro. Caries Research. 2000;**34**(6):498. DOI: 10.1159/000016630

[19] Holmes J. Clinical reversal of root caries using ozone, double-blind, randomised, controlled 18-month trial. Gerodontology. 2003;**20**(2):106-114. DOI: 10.1111/j.1741-2358.2003.00106.x

[20] Samuel SR, Dorai S, Khatri SG, Patil ST. Effect of ozone to remineralize initial enamel caries: In situ study. Clinical Oral Investigations. 2016 Jun;**20**(5):1109-1113. DOI: 10.1007/s00784-016-1710-x

[21] Polydorou O, Pelz K, Hahn P. Antibacterial effect of an ozone device and its comparison with two dentin-bonding systems. European Journal of Oral Sciences. 2006;**114**(4):349-353. DOI: 10.1111/j.1600-0722.2006.00363.x

[22] Rickard GD, Richardson R, Johnson T, McColl D, Hooper L. Ozone therapy for the treatment of dental caries. Cochrane Database of Systematic Reviews. 2004;**3**:CD004153. DOI: 10.1002/14651858.CD004153.pub2

[23] Holmes J, Daley T. Sensitivity & Cracked Teeth Treatment With Ozone. 2003. Available from: http://www.the-o-zone.cc/book/bookch/ch25.html [Accessed: December 30, 2017]

[24] Deepa T, Pooja S. Ozone therapy in conservative dentistry and endodontics: An overview. Indian Journal of Stomatology. 2012;**3**:165-169

[25] Tessier J, Rodriguez PN, Lifshitz F, Friedman SM, Lanata EJ. The use of ozone to lighten teeth. An experimental study. Acta Odontológica Latinoamericana. 2010;**23**(2):84-89. PMID: 21053679

[26] Avaible from: http://www.wh.com/en_global/dental-newsroom/reportsandstudies/new-article/00228/ [Accessed: December 30, 2017]

[27] Nagayoshi M, Fukuizumi T, Kitamura C, Yano J, Terashita M, Nishihara T. Efficacy of ozone on survival and permeability of oral microorganisms. Oral Microbiology and Immunology. 2004;**19**(4):240-246. DOI: 10.1111/j.1399-302X.2004.00146.x

[28] Huth KC, Quirling M, Lenzke S, Paschos E, Kamereck K, Brand K, et al. Effectiveness of ozone against periodontal pathogenic microorganisms. European Journal of Oral Sciences. 2011;**119**(3):204-210. DOI: 10.1111/j.1600-0722.2011.00825.x

[29] Ramzy MI, Gomaa HE, Mostafa MI. Management of aggressive periodontitis using ozonized water. Egypt. Med. J. N R C. 2005;**6**(1):229-245. Avaible from: http://www.the-o-zone.cc/research/abstracts/007.pdf [Accessed: April 06, 2018]

[30] Dodwad VGS, Kumar K, Sethi M, Masamatti S. Changing paradigm in pocket therapy-ozone versus conventional irrigation. Journal of Public Health Dentistry. 2011;**2**(7):7-12. Avaible from: http://journalgateway.com/ijphd/article/view/437/768 [Accessed: December 30, 2017]

[31] Eltas A, Yavuzer D. Dişeti enflamasyonun tedavisinde gaz ozonun klinik etkilerinin değerlendirilmesi. İnönü Üniversitesi Sağlık Bilimleri Dergisi. 2012;**2**:29-33. Article in Turkish

[32] Yilmaz S, Algan S, Gursoy H, Noyan U, Kuru BE, Kadir T. Evaluation of the clinical and antimicrobial effects of the Er:YAG laser or topical gaseous ozone as adjuncts to initial periodontal therapy. Photomedicine and Laser Surgery. 2013;**31**(6):293-298. DOI: 10.1089/pho.2012.3379

[33] Karapetian VE, Neugebauer J, Clausnitzer CE, Zoller JE. Avaible from: http://www.helbo.de/fileadmin/docs/wissenschaft/poster_karapetian_0304.pdf [Accessed: December 30, 2017]

[34] Hauser-Gerspach I, Vadaszan J, Deronjic I, Gass C, Meyer J, Dard M, et al. Influence of gaseous ozone in peri-implantitis: Bactericidal efficacy and cellular response. An in vitro study using titanium and zirconia. Clinical Oral Investigations. 2012;**16**(4):1049-1059. DOI: 10.1007/s00784-011-0603-2

[35] Ozdemir H, Toker H, Balci H, Ozer H. Effect of ozone therapy on autogenous bone graft healing in calvarial defects: A histologic and histometric study in rats. Journal of Periodontal Research. 2013;**48**(6):722-726. DOI: 10.1111/jre.12060

[36] Kazancioglu HO, Ezirganli S, Aydin MS. Effects of laser and ozone therapies on bone healing in the calvarial defects. The Journal of Craniofacial Surgery. 2013;**24**(6):2141-2146. DOI: 10.1097/SCS.0b013e3182a244ae

[37] Kan B, Sencimen M, Bayar GR, Korkusuz P, Coskun AT, Korkmaz A, et al. Histomorphometric and microtomographic evaluation of the effects of hyperbaric oxygen and systemic ozone, used alone and in combination, on calvarial defect healing in rats. Journal of Oral and Maxillofacial Surgery. 2015;**73**(6):1231. DOI: 10. 10.1016/j.joms.2015.02.018

[38] Alpan AL, Toker H, Ozer H. Ozone therapy enhances osseous healing in rats with diabetes with calvarial defects: A morphometric and immunohistochemical study. Journal of Periodontology. 2016;**87**(8):982-989. DOI: 10.1902/jop.2016.160009

[39] Stubinger S, Sader R, Filippi A. The use of ozone in dentistry and maxillofacial surgery: A review. Quintessence International. 2006;**37**(5):353-359. PMID: 16683682

[40] Kazancioglu HO, Kurklu E, Ezirganli S. Effects of ozone therapy on pain, swelling, and trismus following third molar surgery. International Journal of Oral and Maxillofacial Surgery. 2014;**43**(5):644-648. DOI: 10.1016/j.ijom.2013.11.006

[41] Ahmedi J, Ahmedi E, Sejfija O, Agani Z, Hamiti V. Efficiency of gaseous ozone in reducing the development of dry socket following surgical third molar extraction. European Journal of Dentistry. 2016;**10**(3):381-385. DOI: 10.4103/1305-7456.184168

[42] Yucesoy T, Kutuk N, Canpolat DG, Alkan A. Comparison of ozone and photo-biomodulation therapies on mental nerve injury in rats. Journal of Oral and Maxillofacial Surgery. 2017;**75**(11):2323-2332. DOI: 10.1016/j.joms.2017.04.016

[43] Vescovi P, Nammour S. Bisphosphonate-related osteonecrosis of the jaw (BRONJ) therapy. A critical review. Minerva Stomatologica. 2010;**59**(4):181-203; 204-213. PMID: 20360666

[44] Agapov VS, Shulakov VV, Fomchenkov NA. Ozone therapy of chronic mandibular osteomyelitis. Stomatologiia (Mosk). 2001;**80**(5):14-17. PMID: 11696944

[45] Agrillo A, Petrucci MT, Tedaldi M, Mustazza MC, Marino SM, Gallucci C, et al. New therapeutic protocol in the treatment of avascular necrosis of the jaws. The Journal of Craniofacial Surgery. 2006;**17**(6):1080-1083. DOI: 10.1097/01.scs.0000249350.59096.d0

[46] Agrillo A, Ungari C, Filiaci F, Priore P, Iannetti G. Ozone therapy in the treatment of avascular bisphosphonate-related jaw osteonecrosis. The Journal of Craniofacial Surgery. 2007;**18**(5):1071-1075. DOI: 10.1097/scs.0b013e31857261f

[47] Batinjan G, Filipovic Zore I, Vuletic M, Rupic I. The use of ozone in the prevention of osteoradionecrosis of the jaw. Saudi Medical Journal. 2014;**35**(10):1260-1263. PMCID: PMC4362119

[48] Akdeniz SS, Beyler E, Korkmaz Y, Yurtcu E, Ates U, Araz K, et al. The effects of ozone application on genotoxic damage and wound healing in bisphosphonate-applied human gingival fibroblast cells. Clinical Oral Investigations. 2017:11. DOI: 10.1007/s00784-017-2163-6

[49] Dogan R, Hafiz AM, Kiziltan HS, Yenigun A, Buyukpinarbaslili N, Eris AH, et al. Effectiveness of radiotherapy+ozone on tumoral tissue and survival in tongue cancer rat model. Auris, Nasus, Larynx. 2017;**45**:5. DOI: 10.1016/j.anl.2017.03.017

[50] Arita M, Nagayoshi M, Fukuizumi T, Okinaga T, Masumi S, Morikawa M, et al. Microbicidal efficacy of ozonated water against *Candida albicans* adhering to acrylic denture plates. Oral Microbiology and Immunology. 2005;**20**(4):206-210. DOI: 10.1111/j.1399-302X.2005.00213.x

[51] Oizumi M, Suzuki T, Uchida M, Furuya J, Okamoto Y. In vitro testing of a denture cleaning method using ozone. Journal of Medical and Dental Sciences. 1998;**45**(2):135-139. PMID:11186199

[52] Kumar T, Arora N, Puri G, Aravinda K, Dixit A, Jatti D. Efficacy of ozonized olive oil in the management of oral lesions and conditions: A clinical trial. Contemporary Clinical Dentistry. 2016;**7**(1):51-54. DOI: 10.4103/0976-237X.177097

[53] Howard JA. Temporomandibular joint disorders in children. Dental Clinics of North America. 2013;**57**(1):99-127. DOI: 10.1016/j.cden.2012.10.001

[54] Celakil T, Muric A, Gokcen Roehlig B, Evlioglu G, Keskin H. Effect of high-frequency bio-oxidative ozone therapy for masticatory muscle pain: A double-blind randomised clinical trial. Journal of Oral Rehabilitation. 2017;**44**(6):442-451. DOI: 10.1111/joor.12506

[55] Celakil T, Muric A, Gokcen Roehlig B, Evlioglu G. Management of pain in TMD patients: Bio-oxidative ozone therapy versus occlusal splints. Cranio. The Journal of Craniomandibular & Sleep Practice. 2017:1-9. ISSN: 0886-9634 (Print) 2151-0903 (Online) Journal homepage: http://www.tandfonline.com/loi/ycra20 doi:10.1080/08869634.2017.1389506

[56] Nagayoshi M, Kitamura C, Fukuizumi T, Nishihara T, Terashita M. Antimicrobial effect of ozonated water on bacteria invading dentinal tubules. Journal of Endodontia. 2004;**30**(11):778-781. PMID: 15505509

[57] Hems RS, Gulabivala K, Ng YL, Ready D, Spratt DA. An in vitro evaluation of the ability of ozone to kill a strain of *Enterococcus faecalis*. International Endodontic Journal. 2005;**38**(1):22-29. DOI: 10.1111/j.1365-2591.2004.00891.x

[58] Estrela C, Estrela CR, Decurcio DA, Hollanda AC, Silva JA. Antimicrobial efficacy of ozonated water, gaseous ozone, sodium hypochlorite and chlorhexidine in infected human root canals. International Endodontic Journal. 2007;**40**(2):85-93. DOI: 10.1111/j.1365-2591.2006.01185.x

[59] Siqueira JF Jr, Rôças IN, Cardoso CC, Macedo SB, Lopes HP. Antibacterial effects of a new medicament—the ozonized oil—compared to calcium hydroxide pastes (original article in Portuguese). Revista Brasileira de Odontologia. 2000;**57**:252-256

[60] Reddy SARN, Dinapadu S, Reddy M, Pasari S. Role of ozone therapy in minimal intervention dentistry and endodontics—A review. Journal of International Oral Health. 2013;**5**(3):102-108. PMCID: PMC3769872

[61] Good M, El KI, Hussey DL. Endodontic 'solutions.' Part 1: A literature review on the use of endodontic lubricants, irrigants and medicaments. Dental Update. 2012;**39**(4):239-240; 42-44, 46. PMID: 22774686

[62] Ghobashy SA, El-Tokhey HM. In vivo study of the effectiveness of ozonized olive oil gel on inhibiting enamel demineralization during orthodontic treatment. Journal of American Science. 2012;**8**(10):657-666. Avaible from: http://www.jofamericanscience.org/journals/am-sci/am0810/. [Accessed: December 30, 2017]

[63] Cehreli SB, Guzey A, Arhun N, Cetinsahin A, Unver B. The effects of prophylactic ozone pretreatment of enamel on shear bond strength of orthodontic brackets bonded with total or self-etch adhesive systems. European Journal of Dentistry. 2010;**4**(4):367-373. PMCID: PMC2948738

[64] Grootveld M, Baysan A, Sidiiqui N, Sim J, Silwood C, Lynch E. History of clinical applications of ozone. In: Lynch E, editor. Ozone: The Revolution in Dentistry. London: Quintessence Publishing; 2004. pp. 23-30

[65] Bezirtzoglou E, Cretoiu SM, Moldoveanu M, Alexopoulos A, Lazar V, Nakou M. A quantitative approach to the effectiveness of ozone against microbiota organisms colonizing toothbrushes. Journal of Dentistry. 2008;**36**(8):600-605. DOI: 10.1016/j.jdent.2008.04.007

[66] Baysan A, Lynch E. Clinical reversal of root caries using ozone: 6-month results. American Journal of Dentistry. 2007;**20**(4):203-208. PMID: 17907479

[67] Gojanur S. Ozone Therapy—An Alternative to Conventional Therapy in Pediatric Dentistry. 2016. Avaible from: http://wwwdentalnewsweekcom/?p=991 [Accessed: December 30, 2017]

[68] Margolis HC, Moreno EC, Murphy BJ. Importance of high pKA acids in cariogenic potential of plaque. Journal of Dental Research. 1985;**64**(5):786-792. DOI: 10.1177/00220345850640050101

[69] Jayarajan J, Janardhanam P, Jayakumar P. Efficacy of CPP-ACP and CPP-ACPF on enamel remineralization—An in vitro study using scanning electron microscope and DIAGNOdent. Indian Journal of Dental Research. 2011;**22**(1):77-82. DOI: 10.4103/0970-9290.80001

[70] Brazzelli M, McKenzie L, Fielding S, Fraser C, Clarkson J, Kilonzo M, et al. Systematic review of the effectiveness and cost-effectiveness of HealOzone for the treatment of occlusal pit/fissure caries and root caries. Health Technology Assessment. 2006;**10**(16):1-6 PMID: 16707073

[71] Huth KC, Paschos E, Brand K, Hickel R. Effect of ozone on non-cavitated fissure carious lesions in permanent molars. A controlled prospective clinical study. American Journal of Dentistry. 2005;**18**(4):223-228. PMID: 16296426

[72] Dahnhardt JE, Jaeggi T, Lussi A. Treating open carious lesions in anxious children with ozone. A prospective controlled clinical study. American Journal of Dentistry. 2006;**19**(5):267-270. PMID: 17073201